CULTIVATING GENDER

GENDERING ASIA

A Series on Gender Intersections

Gendering Asia is a well-established and exciting series addressing the ways in which power and constructions of gender, sex, sexuality and the body intersect with one another and pervade contemporary Asian societies. The series invites discussion of how people shape their identities as females or males and, at the same time, become shaped by the very societies in which they live. The series is concerned with the region as a whole in order to capture the wide range of understandings and practices that are found in East, Southeast and South Asian societies with respect to gendered roles and relations in various social, political, religious, and economic contexts. As a multidisciplinary series, *Gendering Asia* explores theoretical, empirical and methodological issues in the social sciences.

Series Editors: Wil Burghoorn, Gothenburg University; Monica Lindberg Falk, Lund University; Cecilia Milwertz, NIAS; and Pauline Stoltz, Aalborg University (contact details at: http://www.niaspress.dk).

1. *Working and Mothering in Asia. Images, Ideologies and Identities,* edited by Theresa W. Devasahayam and Brenda S.A. Yeoh
2. *Making Fields of Merit. Buddhist Female Ascetics and Gendered Orders in Thailand,* by Monica Lindberg Falk
3. *Gender Politics in Asia. Women Manoeuvring within Dominant Gender Orders,* edited by Wil Burghoorn, Kazuki Iwanaga, Cecilia Milwertz and Qi Wang
4. *Lost Goddesses. The Denial of Female Power in Cambodian History,* by Trudy Jacobsen
5 *Gendered Inequalities in Asia. Configuring, Contesting and Recognizing Women and Men,* edited by Helle Rydstrøm
6. *Submitting to God. Women and Islam in Urban Malaysia,* by Sylva Frisk
7. *The Authority of Influence. Women and Power in Burmese History,* by Jessica Harriden
8. *Beyond the Singapore Girl. Discourses of Gender and Nation in Singapore,* by Chris Hudson
9. *Vietnam's New Middle Classes: Gender, Career, City* by Catherine Earl
10. *Gendered Entanglements: Revisiting Gender in Rapidly Changing Asia,* edited by Ragnhild Lund, Philippe Doneys and Bernadette P. Resurrección
11. *Queer/Tongzhi China: New Perspectives on Research, Activism and Media Cultures,* edited by Elisabeth L. Engebretsen and William F. Schroeder (with Hongwei Bao)
12. *Cultivating Gender: Meanings of Place and Work in Rural Vietnam,* by Cecilia Bergstedt

NIAS Press is the autonomous publishing arm of NIAS – Nordic Institute of Asian Studies, a research institute located at the University of Copenhagen. NIAS is partially funded by the governments of Denmark, Finland, Iceland, Norway and Sweden via the Nordic Council of Ministers, and works to encourage and support Asian studies in the Nordic countries. In so doing, NIAS has been publishing books since 1969, with more than two hundred titles produced in the past few years.

UNIVERSITY OF COPENHAGEN

norden
Nordic Council of Ministers

CULTIVATING GENDER

Meanings of Place and Work in Rural Vietnam

Cecilia Bergstedt

NIAS PRESS

Nordic Institute of Asian Studies
Gendering Asia series, no. 12

First published in 2016 by NIAS Press
NIAS – Nordic Institute of Asian Studies
Øster Farimagsgade 5, 1353 Copenhagen K, Denmark
Tel: +45 3532 9501 • Fax: +45 3532 9549
E-mail: books@nias.ku.dk • Online: www.niaspress.dk

A CIP catalogue record for this book is available from the British Library

ISBN: 978-87-7694-179-6 (hbk)
ISBN: 978-87-7694-180-2 (pbk)

Typeset in Arno Pro 12/14.4
Typesetting by Lene Jakobsen
Cover concept/illustration: Sara Forsberg

Printed and bound in Great Britain
by Marston Book Services Limited, Oxfordshire

Contents

Preface *vii*
Introduction 1
1. 'Women', 'peasants' and land – ideological and historical contexts 15
2. The village of Lang Xanh 47
3. Continuity and change – gender and land access 83
4. 'Big work' and 'small work' – gender and labour division 129
5. Cultivating community – gender and labour organisation 169
Conclusions 195
References 207
Index 217

List of Figures

1. *The Harvesters* by Pieter Bruegel the Elder 7

2. Old women on their way to the pagoda 81

3. The ancestor altar in my host family's house 88

4. Ritual offerings 93

5. Summer transplanting 131

6. Harvesting the first rice crop of the year 135

All photographs (figs 2–6) by the author.

Preface

This is a book about the work and everyday life of the villagers in Lang Xanh, and it required the collaboration of many persons. A couple of weeks before the fieldwork was about to commence, Phuong, my interpreter, and I visited the People's Committee in the commune to discuss the practical arrangements of our stay. I had all along imagined and hoped that we would find a family to stay with, but at first, the persons from the People's Committee held another opinion. Luckily, after I had nearly exhausted my reservoir of reasons why I preferred a family's home to commuting back and forth from the nearest town, or staying at the healthcare station, a family was suggested as a possible host for us. After visits and negotiations, it was decided that we could come and stay with them.

The family consisted of Minh, a schoolteacher, Lan, a fulltime farmer, their teenage daughter, and Minh's widowed mother. Minh and Lan also had two older sons who had left the home to work in Hanoi and Ho Chi Minh City. In their absence Phuong and I could occupy one of the large beds in the main room, while Minh and Lan slept in the other. Minh's mother and the daughter shared the smaller room. This came to be the arrangement for the entire year of the fieldwork.

Because we lived with a family, every day and almost every hour was fieldwork time. This was what I had hoped for. I wanted to be present and participate as much as possible in everyday life. Therefore, it was a great privilege to be able to live with the same family and be accompanied by Phuong as my interpreter throughout the whole stay. My host family's house was my home and point of departure and the place from where I connected with the other persons, households and places of the village. Naturally, I came to know my host family best and some of their friends and relatives became my first acquaintances. With them I almost daily discussed all kinds of matters and, through them, I got as much of an inside view as was possible. As my social network gradually expanded, I met persons who became invaluable to my study but who also warmed my heart because their kind treatment made my time in the village enjoyable. Through these encounters with the people in Lang Xanh, I learnt not only about their lives but also quite a bit more about my own.

This book fully depends on the cooperation and patience that the villagers of Lang Xanh showed. I am deeply grateful for all the time they spent with me and for sharing their thoughts and places with me. I am also most thankful to Luong Thi Kim Phuong, who was my interpreter, but also my companion and partner in discussions about a myriad of topics that concerns our human condition. She contributed greatly to my fieldwork, and her family sometimes offered appreciated little pauses from fieldwork.

This book, which started out as a doctoral project, was made possible through the founding by Sida/SAREC. I was also the grateful recipient of a grant from the Swedish Foundation for International Cooperation in Research and Higher Education (STINT), which made it possible for me to spend an academic year at the School of Oriental and African Studies (SOAS) in London. And I thank Adlerbertska Stiftelsen for financial support. The Centre for East and South-East Asian Studies at Lund University, with Professor Roger Greatrex as the Director, granted me a much appreciated Short-term Research Fellowship. I am also indebted to the Nordic Institute for Asian Studies (NIAS) in Copenhagen, for giving me a NIAS SUPRA Nordic scholarship.

Professor Khong Dien, the Director of the Institute of Anthropology at the National Centre for Social Sciences and Humanities in Hanoi at the time of this fieldwork, and his staff, provided me with vital support and assistance, not least with acquiring the permissions necessary for conducting fieldwork in Vietnam.

Through my fieldworks in Vietnam, I was lucky to meet with Thao Do Jörgensen. She started out as an excellent interpreter and quickly turned into a dear friend. For the first couple of weeks of fieldwork for this project, she accompanied me to Lang Xanh, and she edited the Vietnamese words for this book. A special thanks goes to the entire Do family who have shown me great hospitality and provided a home away from home on my visits to Hanoi. If it was not for Dr Bent D. Jörgensen,

I might never have come to know *em* Thao, and it was through my involvements in his projects that I first came to Phu Tho province, where Lang Xanh is situated.

As a PhD candidate, I enjoyed the company of several valuable persons and, not least, the collective of doctoral candidates in Social Anthropology at the School of Global Studies, University of Gothenburg. I am happy to have met each one of you. First and foremost I would like to warmly thank Dr Wil Burghoorn, who was my supervisor, for all the years of encouragement and support. I feel privileged to know her. Professor Åsa Boholm contributed greatly with her comments when the text was still a manuscript for my thesis. Professor Kaj Århem has also provided useful comments on one of the chapters. I would also like to mention Associate Professor Claes Corlin, who pointed me in the direction of Vietnam by enabling me, as a Master student, to conduct my first two periods of fieldwork there. For their reading and commenting on early versions of some chapters of this work, I am extra thankful to Dr Nina Gren and Dr Kristina Nässén. Karin Larsson Wentz and Judith Crawford did indispensible work with language editing on an earlier version of the manuscript. My thoughts of appreciation includes Sara Forsberg, whose artistic talent resulted in the cover of this book.

At NIAS Press, I am grateful to Gerald Jackson, Editor-in-Chief, for showing confidence in this book and to the editors for comments and language improvements. The support of the Gendering Asia series editors was also very important. In addition, I much appreciated the opinions and commentaries from the three anonymous peer reviewers.

Last, but never least, I want to express my sincere appreciation for those persons – family and friends – who have been with me through the process of making this book. While I have had little to offer you, I have enjoyed your understanding, care and consideration. This holds especially true for Peter, whom I cannot thank enough.

Introduction

'The husband ploughs, the wife transplants, the buffalo harrows' (*Chong cay, vo cay, con trau di bua*).

This saying had survived the communist revolution, the land reforms, and the market-oriented household farming of today's Lang Xanh, a smallish rural village in the northern part of Vietnam where I had come to do my fieldwork. In some respect, this trinity still constituted the essence of farming life – a simple, self-evident, and effective division of labour that was indisputable as long as the means for other ways of life and support were not available. 'The land tells us what to do and all we have to do is follow', as Viet, a male farmer, described his perception of the small-scale and manual farming the people of Lang Xanh engaged in. This 'natural' land-generated perception of life and labour was also reflected in the often-used phrase: men do 'big work'(*viec lon*) and women do 'small work' (*viec nho*).[1] This implied, for example, that ploughing was considered as 'big', while transplanting fell into the category of 'small' work.

This gendered division of farm labour tasks had spatial implications in that women and men frequently performed their gender-specific tasks separated from each other in place and quite often also in time. In this sense, the practices of farming actually contributed to create places that harboured gender-particular intentions and implication for the dwellers. The meanings of and access to particular places were, as this book will show, closely bound up with ideas about gender.

1 The English distinctions between work and labour do not really have their equivalents in Vietnamese. *Cong viec* is the noun for work, or job, and *lam viec* is the verbal form, i.e. to work or to labour. I use work and labour more or less interchangeably, and not as indicators of shifts in meaning or value (c.f. Arendt 1958).

It was the changing seasons of the farming year – and especially the cultivation of wet rice – that shaped much of the work and the environment of the village Lang Xanh. At the peaks, groups of 10 to 15 women could be seen bending, side by side, either transplanting the bright green rice seedlings into the muddy water filled fields, or cutting the ripe rice and tying the stalks into sheaves. Most of them would be wearing dark trousers, conical hats and, if it was sunny, a scarf or a mask that protected their faces from the sun's tanning rays. At a first glance, it was the bright colours of the women's shirts that distinguished them from each other – at least for a stranger, like myself. Often the women were chatting, joking, and exchanging comments and opinions among themselves. Between harvesting and the next transplanting, the fields were usually harrowed, or sometimes hoed, and then ploughed by a solitary man and his buffalo. There was a special 'language' that was used for instructing the buffalos to start, stop, or turn right or left, and these commands could be heard echoing around the area from morning to late afternoon when these men were at their busiest. During off-peak seasons, male or more often female farmers were in motion in and around the fields as they paid regular visits to check on the water level, making sure that the banks surrounding each field regulating the water were intact, or just seeing how things were going.

Especially at the busiest times of the farming year, the tasks were generally divided according to an ideal order where gender played a significant role. On household or individual level, deviations and exceptions were sometimes made for a multitude of reasons, but there existed an established general division of farm labour.

'If you were a man, all the women would want to marry you', one woman told me after I had spent a month or so in Lang Xanh. We were a group of women who had gathered in the shade of a large tree that grew between the fields and the garden of the village leader's house. There we ate sweet and sour starfruit while some of the women enjoyed a rest from weeding the rice fields. When I heard her remark, I was not sure what she meant. But after some elucidatory comments, I understood that what had triggered her remark was how I moved around in the village, and in the world. She and other villagers had noticed my habit of walking the paths of the village and the surrounding fields and chatting or discussing various things with people I met on the way. If someone invited me to come to their home I would go and visit them and maybe have a cup of tea or a snack while we conversed about this or that. My 'socialising' included virtually everyone across the social spectrum. This breadth reflected the fact that I had not only been to the capital, Hanoi, and the local surroundings, but also to other continents as well. The women explained to me that this was a man's way of moving and occupying places. To be able – at least ostensibly – to wander around purposelessly for no apparent reason and with no destination and idly fall into conversation with anyone and accept an invitation without careful estimations about what people might think was a man's way of moving around. A woman would find it difficult to be able to leave home without risking to be accused of neglecting one's responsibility for household work. To be seen 'hanging around' outdoors without working or being on the way to or from some engagement was predominantly a male privilege. I had behaved more like a man – going around, wanting to get acquainted with everyone and, eventually, I would go back to my western country to resume my work at my university. This behaviour befitted a man with certain social skills and a relatively comfortable life, a man that women might want to marry. However, it was not really proper behaviour for a woman, and especially not, the village women let me understand, a woman who should have been a wife and a mother long time ago according to my age (36 at the time). My way of approaching this place was odd for a woman and in their eyes I had put my whole future as a woman at risk by leaving my home.

Some time after, at a dinner one evening with my host family and one of their male friends, we elaborated on the popular and often thought-provoking topic of advantages and disadvantages being born a man or a woman. Among the more jokingly comments and suggestions, our guest's argument for why he would never want to be a woman stood out: 'Women cannot move around'. He stated this with some emphasis even though it could easily be proven inapplicable to some of the women in Lang Xanh. However, his idea of women's constrained mobility was indicative. Later on, this opinion was further enforced by women who told me about questions and accusations from suspicious husbands that awaited wives who were not at home, or had returned back later than expected, without an acceptable explanation for their whereabouts. Wandering about was not a matter of course for the female inhabitants of this place. Outside of the home, one of the most appropriate places to be for women of working age was in the fields. For many of the women, this was one of the places where they spent most of their time, sometimes alone and during the peaks of the farming year, in the company of others. In comparison with the increasing necessity of – and in some cases opportunity to – work 'outside' of farming and the village community, the daily and near-home labour that mostly women performed in the fields was comparatively safely set 'inside' the moral community. As we shall see, the women of Lang Xanh also made use of this 'safe' place when they cultivated their female traits.

This book investigates the practices through which the inhabitants of Lang Xanh engaged in processes of gendering while they simultaneously cultivated their land. The aim is to explore how their ideas and perceptions of femininity and masculinity were manifested in some of the situations when they were interacting with farmland and residential land respectively. Meeting the farmers in the village, I was incited to investigate how gender become naturalised through the integration of practices and places, so that the actions and their meanings seem to emanate from the gendered bodies in a 'natural' manner – as if any other alternative would seem awkward and irrational. I will argue that such naturalised practices, like the division of labour for example, imply that the bodies of the farmers have been directed, or oriented, toward places and activities in gender-specific ways.

One of the reasons why farmland came to be the main focus of this gender study was the ubiquity and 'naturalness' of land and farming – in this case predominantly wet-rice cultivation – and the matter-of-factly manners in which the gendered labour division was presented by the villagers. This spurred me to investigate how female and male capacities, which were seemingly taken for granted, came to be meaningful in the particular context of farmland. Following along this line of investigation, my attention was turned to the day-to-day practices and events, and not so much to the spectacular or the eye-catching occurrences associated with for example ceremonial or extraordinary episodes in life. In addition to farmland, women's and men's access and relation to residential land will be explored. This will bring out complementary as well as contrasting perspectives on the relationship between perceptions of gender and land. By putting the daily routines of land-related practices at the centre of attention, the processes of gendering that occur should emerge in the light of the themes of each ethnographic chapter. With everyday life in focus, dealings with land and farming can be understood as meaningful and influential on how self and others are understood as gendered beings. Moreover, the way people relate to and perform land-oriented practices can be informed by their gendered situations.

Although ideas and practices concerning femininity and masculinity largely come into being in relation to each other, there is a predominance of women and perceptions of femininity in this book. This is so for several reasons. My first encounter with gender and land issues was through field studies of northern Vietnamese female-headed households and these women's situation in and experiences of allocation of farmland. This was followed by other field studies focusing on decision-making and processes of democratisation on a village level, mainly from a gender perspective. It became increasingly clear to me how much time and effort Vietnamese women put into their farm work, how different the conditions often were for men and women in rural areas and, as a consequence, how important gender was for daily practices regarding arable land and access to residential land and housing. Another factor that influences the balance between the presence of men and women in this book is the fact that women performed many of the chores involved in farming. The feminisation of farming was rather widespread, especially in the daily maintenance of the fields. I shall elaborate on the reasons and consequences of this in the following chapters. However, an apparent effect was that fewer men were out in the fields and thus available to discuss the everyday efforts of farming. Last but not least, the fact that I am a woman who tried to be as present in the fields as possible made my encounters with women both more frequent and more intimate. Nonetheless, men were definitely part of farming and some of them became engaged and valuable participants in the work with this book. Eventually I learned that men's absence was as meaningful and important as their presence in relation to farm labour and masculinity. This will be elaborated further later on.

Even if the focus of this book and the data collected is not of a particularly sensitive kind – although issues of for example land access can be very problematic and have on several occasions turned violent in different places in the country – I have decided to give the persons who appear in the text other names than their real ones. Because the People's Committee of the commune was situated in the village, making it very easy to identify the village, I have also decided not to disclose the name of the commune and to give the village a fictive name. In some way, this is a pity, since the name of the commune is quite beautiful and the village name is almost funny in its matter-of-factly descriptive nature. Nonetheless, Lang Xanh is the name I will call the village. This means 'the green village', which I find quite nice and unobtrusive in its simplicity.

In sum, this book wants to explore how social and spatial aspects – as they emerge through everyday practices and perceptions – influence the understanding of individuals, practices and places as gendered. A female or male body has its own particular way to relate to and move in the environment. Also, spatial separation of women and men's labour and duties, for instance, reinforce perceptions of distinct male and female capacities (Spain 1992: 85–86). Therefore, I will try to show that the residential land and the rice fields of Lang Xanh were integral parts of the process of gendering and are thus in themselves gendered places. From such a perspective, the fields were not just neutral ground where rice was cultivated, but also places where men and women were cultivated.

Gendered dwellers

In order to investigate the ways in which Lang Xanh peasants' engagement with the land shapes them into meaningful and acknowledged female and male farmers, a phenomenologically influenced idea of people as dwellers in the world will provide a vantage point. A 'dwelling perspective', which I predominantly borrow from the anthropologist Tim Ingold, builds on the assumption that persons and their environment form a mutually informing and constructive relationship, primarily through people's practical involvements with their environment (Ingold 2000: 153, 154, 191). The shaping of the environment as well as of the dwellers emerge from people's movements and skilled practices in particular places. This reciprocal engagement is a 'movement wherein forms themselves are generated' (ibid.: 193). In this perspective, all our practices get their meaning in relation to other persons, other tasks, and other places (ibid.: 195). With this in mind, studying people's interactions with land, we can learn how existence is constituted in particular and meaningful ways.

Figure 1: *The Harvesters* by Pieter Bruegel the Elder

To dwell in the world is ultimately a social activity because we are not only aware of the places where we act, but we are also aware of the presence of others in our social environment (ibid.: 196). Ingold analyses the painting *The Harvesters* (1565) by Pieter Bruegel the Elder in order to illustrate that the landscape, just like human life and activity, is temporal and experienced in an immediate and relational way. He prefers this view of human–environment relations to seeing the world as covered with layers of meaning that can be interpreted independently from the persons who inhabit the places and the activities they perform there. A relational view, where places, persons and their activities are mutually constitutive of each other, is also preferable to perceiving the environment as a persistent entity upon which humans enact their lives (ibid.: 207).

Ingold vividly takes the reader through the landscape of the painting, drawing our attention to how everything, depending on temporal perspective, is moving and changing. The passing of time is reflected differently in a tree or a church compared to crop. Even relatively durable attributes, like the large pear tree or the church in the valley, form organic parts of the environment. Their specific features contribute to making the environment turn into particular places with certain, place specific, characteristics. Ingold makes us aware that we experience the landscape through our muscles, moving along paths, up hills, down tracks, eyeing the valleys, working the fields. This leave marks in the landscape, shaping it at the same time as our muscles are toned through our daily movement in a place over generations. The reciprocity between persons and environment is unmistakeably clear when it comes to the vast wheat field where the harvesters in the painting are working and resting. The farm workers are continuously engaging with the field and are drawn together by the cycles of the farming seasons, their division of labour, coordination of tasks and working pace. Neither the viewer of the painting nor the harvesters occupying the landscape can be separated from any of the components in the scenery. Our minds and perceptions are *with* the environment, and *with* our bodies in the landscape. We cannot operate outside of this, observing the surroundings from a detached and bodiless position (ibid.: 201–207).

When Inglod's enquiry turns to the people who inhabit the painting, predominated by those engaged in harvesting the rich crop of wheat, or taking a break to eat and rest, he emphasis how they and the landscape are drawn together in activities responsive both to the conditions of the environment and to their fellow harvesters. People are assigned to different tasks but never act in isolation. According to Ingold's interpretation, they 'work in unison, achieving a dance-like harmony in their rhythmic movements' (ibid.: 207). What he does not notice is that it is only men who cut the wheat and tie it into sheaves, and that only women carry the prepared sheaves away from the field. The harmonious dance of harvesting is apparently incorporated into the dwellers either as male cutters of the wheat or female carriers of sheaves. In Bruegel's painting, the complementarity of acting bodies and landscape has generated gendered dwellers who engage with their environment and their fellow workers according to their gender-specific skills and tasks. Hence, the world that the dwellers in engagement with the landscape generates, indicates an environment wherein male and female farm labourers move and act in accordance with gender-specific knowledge and capacity.

What often seems to happen when people go about their daily chores in their well-known home environment is that practices are taken for granted and appear as 'natural'. In other words, when gendered conditions emerge in the form of people's everyday engagements with their surroundings, they often escape 'the level of discourse' (Bourdieu 1977: 87) in such a way as the gendered division of labour did for Ingold. This book explores how, in Lang Xanh, the gendering habits of farm labour emerge through practical bodily enactment in particular places. In this process, persons and places become ordered according to a logic that generates differences between feminine and masculine bodies that appear as self-evident in their 'natural' setting. Much of human actions are performed in accordance with the forms and qualities of the land without consciously reflecting about it; we take what we do and how we interact with our environment for granted because our practices and behaviour serves our purposes and intensions (Crossley 2001: 89–90, 100; Jackson 1989: 130).

Gender is an intrinsic part of how we move and act in the world and is one of the factors that influence our conditions and experiences in certain places. Learning the limits and possibilities of actions in a place is a gendering process in which men and women get a 'feel' for which actions that are appropriate in which places. In her book *Queer Phenomenology*, Sara Ahmed argues that gender is, among other things, a bodily orientation (Ahmed 2006: 60). The shapes that bodies take and the activities that they engage in are the outcome of repetitive actions over time. It is not a coincidence that certain bodies tend to be in particular places doing specific things over and over again. Historical and habitual contexts orient our 'female' or 'male' bodies toward certain places and certain actions, and the constant recurrence makes the work that gendered bodies do appear as 'natural' and 'self-evident'. So, all the hard work and repletion that goes into doing a task effortlessly orient the gendered practitioner into some tasks and some places rather than others. This leads Ahmed to conclude that repetitive work is not neutral, and the seemingly effortless tasks that we engage in actually hide all the hard work behind it. Because the 'naturalness' distracts our attention away from what is actually done by whom – like in the painting of the harvesters – this kind of work tends to hide how we become oriented away from some areas and into others (ibid.: 56–60). Here we find that gendered bodies, work, and place can form a mutually constituting circle wherein, for example, the gendering effect of division of labour can be concealed.

The social and cultural geographer Tim Cresswell insists that 'the social and the spatial are so thoroughly imbued with each other's presence that their analytical separation quickly becomes a misleading exercise' (Cresswell 1996: 11). Meanings and judgements of people's practices are immersed in and dependent on where they take place. What is appropriate in one place can be highly improper somewhere else. As long as our understandings of the relations between actions and places are commonly agreed upon, we get a sense of self-evidence and naturalness (ibid.: 11–22). The outlook of many of the Lang Xanh farmers on their everyday dealings with land reflected such a matter-of-fact attitude.

Cultivating gendered farmers

Implicit throughout the discussions about gender-specific attributes will be the villagers' commonly held perception that a person is basically categorised as either man or woman, based on the genitals of the body, which are identified as either male or female at birth. A Vietnamese woman or man is 'referred to as *being* her or his body' and a person's body is understood to integrate and encompass multifaceted features varying from, e.g., individual fate to emotions and physiology (Rydström 2003: 5). Such an essentialist idea of the body, as Rydström points out, makes oppositions between sex and gender, or between body and mind, quite ineffective as analytical tools for understanding Vietnamese perceptions of the meanings of being a man or a woman. These types of division are not really relevant for them. First and foremost, the essence of the body lies in its generative capacity in relation to a patrilineal kinship system, i.e. whether the body is male – with the ability to pass on its own patrilineal bloodline – or female, and thereby carrying the feminine ability to reproduce not its own, or her father's, but the male procreator's family line (ibid.: 4–5, 92–95). The men and women of Lang Xanh generally shared the basic notion of essentially distinct categories of 'male' and 'female' bodies.

To some extent, the gendering structures in Lang Xanh were all encompassing and compelling to the extent that everyone was categorised and affected whether they liked it or not. Individual women and men were often defined by long-lived structures and had to relate to the consequences of how other people treated them, even though these structures did not always reflect the exact ideas of every person or group in this place (Young 2002: 420, 422). Farm work and access to residential land were areas where gender structures had far reaching consequences for the lives of individual villagers. From one point of view, farm work and land access were domains in which the male and female peasants could pursue – quite consciously – desired types of femininity and masculinity. These domains also provided opportunities for women and men to distinguish themselves in relation to others. However, if we can see women's opportunity to act 'as a capacity for action that historically specific relations of subordination enable and create' (Mahmood 2001: 203), then what might at first look like passive compliance with a gender discourse that works mainly to women's disadvantage, can on closer investigation also turn out to provide a sphere for women's self-realisation (Mahmood 2005: 174). Therefore, I will suggest that the farm labour of Lang Xanh women can be understood as a means to act and to shape themselves as respectable and responsible women within an existing framework of gender norms.

Gender, farmers, and land in Vietnam – a broader perspective

In Vietnam, the categories of 'women', 'peasants'[2] and 'farmland' have been paid a great deal of political, ideological, and scholarly attention, not least since the resistance against the French colonial regime began to take more distinct shapes in the 1920s.

Throughout the period of colonial resistance and the building of a socialist nation, the control of land, and especially farmland, as a means of shifting and executing power – ideologically as well as politically and economically – has been a major issue. The centrality of farmland continued when Vietnam underwent an economic reformation (*Doi moi*) in 1986 that opened up for a market-oriented economy, followed by a land allocation in 1993 when collective farming formally ended (Luong 2010: 166–181; Malarney 2002: 23–51). The necessity to define and manage rural women, agriculture, and land use – not least to handle the needs of shifting political and economic circumstances – remained some of the key issues for state policies in order to achieve desirable pro-

2 In Vietnamese, the word *nong dan* translates into both farmer and peasant. It does not indicate scale or ownership of land.

gress (Le Thi 2001: 109, 115–118; Le Thi Nham Tuyet 2002: 33, 42). Hence, questions concerning land, farming, and women, have all been running through the political movements. Still today, more than 70 per cent of Vietnam's population have their home in the countryside and around 60 per cent of the country's inhabitants have agriculture as their main occupation.[3] Therefore, farmland and farmers – not least women farming the household fields – form large parts of Vietnamese society.

The female farmers of Lang Xanh were acting in an ideological context where socio-cultural, political, and economic circumstances structured their positions and identities. Moreover, as will be dealt with further on, Lang Xanh women could be seen as consciously using and incorporating strategies of feminine virtues and morality through their farm work. In this way, they tapped into the discourses of rural female care and self-sacrificing in which, as Philip Taylor noted, young rural women in the Mekong Delta of south Vietnam were represented as ideal marriage partners not only by others but sometimes also by themselves (Taylor 2007: 29–30). For men working the fields – especially the rice fields in Lang Xanh, as we will see – it was more of a balancing act. They had to juggle their farm work carefully and sometimes even avoid it, in order to prevent a slippage into the female domain and risk losing their masculinity.

The organisation of the book

Chapter 1 offers an historical retrospective that follows the ideological development and the claims that have been made on women, peasants and land in Vietnam from the beginning of the revolutionary movements until the present situation. Here, some of the historical and ideological settings that the farmers in Lang Xanh had to relate to are presented.

Chapter 2 is a presentation of the village, with a focus on land issues. Certain key persons and the functions they represented are introduced and some of the conditions for my stay in the village are considered. Thereafter, Chapter 3 makes a comparison between access to and control of arable land (predominantly rice fields) and residential land from a gender perspective. Women had less opportunity than men to access

3 See: http://web.worldbank.org/wbsite/external/countries/eastasiapacificext/exteapregtoprurdev/0,,contentMDK: 20534368~menuPK: 3127821~pagePK: 34004173~piPK: 34003707~theSitePK: 573964,00.html.

residential land and houses in their own right. Their inability to link generations within the patrilineal family was a barrier to this. Instead, they contributed to the sustenance of male permanency through their practical involvement in the ancestor worship relating to residential land and the family home. In this way, women balanced their transient positions with preserving abilities. Access to farmland was comparably gender equal. This, I argue, was due to the detachment that had taken place between farmers and land through land reforms, collectivisation, and the allocations that the state had implemented.

Chapter 4 shows the interrelationship between farm labour and how women and men acquired gender-specific traits and characteristics. Women's substantial work efforts during times of revolution and warfare have been strongly encouraged and idealised. The existing discourses on women as persistent and able workers made farm labour a realm where Lang Xanh women could pursue their qualities and distinctiveness. Moreover, when women laboured in the fields, they performed work that nourished their husbands' patrilineal family. The daily maintenance of the fields has been incorporated into female morality and this, in turn, made frequent physical presence on the fields an integral part of femininity in Lang Xanh. The kind of female activity that I consider here was not the kind that formed actual resistance against prevailing norms. Instead, it was the kind that made women realise themselves as actors, even if it was within an area, which in certain ways was dominated by men.

Chapter 5 concerns aspects of village communality in relation to farm labour organisation. Most farmers in Lang Xanh depended on each other for a certain level of cooperation. In peak seasons, the farmers had to engage people outside their households to cope with their labour needs. It had become more and more common that some of the households paid for extra labour rather than involve themselves in labour exchange. For some of the villagers, this challenged the idea of village community and loyalty among fellow villagers. Women's engagement in relationships of farm labour exchange could be seen as upholding emotional and relatively undifferentiated social relations, which could even out and somewhat counterbalance the less communal and non-reciprocal aspects of farming and other types of livelihoods. By being present in the village and linking together, first and foremost, other women from different generations and families into corporeal labour groups, women's communal farm labour gave the sense of village community a concrete shape.

The last chapter sums up the main findings and arguments of the book and reflects on how gendering can be understood as processes that integrate presuppositions and experiences of gendered-specific traits into our beings as we go about our daily work in familiar places.

CHAPTER ONE

'Women', 'peasants' and land – ideological and historical contexts

In Vietnam it has been possible, in the span of one lifetime, to experience French colonialism, communist revolution, war against France and, later on, the USA, followed by economic reform politics that let market forces into people's lives on a wider front. All these events have been intertwined with changes in land control and with transformations of the positions of women and peasants[1] in society. During the entire revolutionary era, conscious and explicit normative campaigns concerning women (Pettus 2003: 8–10, Taylor 1999: 24) and peasants (Ngoc-Luu Nguyen 1987: 562–65; White 1983: 74–79) were common. Land reforms were one of the most subversive means used by the communist regime to achieve their goals. As Vietnam moved on from revolution and warfare and into times of reconciliation, nation-building and subsequent economic reformations, these courses of events have

1 'Peasants' can be defined as rural, agricultural, cultivators who have land as central means, and who produce for their own consumption and, often, transfer surplus production to possessors of greater economic and political power (Ngoc-Luu Nguyen 1987: 10, Wolf 1966: 3–4). However, peasants can belong to different strata in society depending, for example, on whether they own the land they cultivate, if they rent it from a landholder, or work the land as serfs and a landlord controls all the produce. Peasants can also be landholders who employ farm labourers and may or may not participate in farming themselves (Keesing 1981: 422–423). The categorisation of Vietnamese peasants, especially in the northern part of the country, has played a central role in the country's history and development, and this will be addressed in the present chapter. As a general definition, this book refers to 'peasants' as rural cultivators who till the land on a household basis and who produce mainly for their own consumption. Another important feature of Vietnamese peasants that Ngoc-Luu Nguyen adopts from Teodor Shanin (1971) is the village as a basic societal unit through which the peasantry relate to other non-rural categories of society and the state (Ngoc-Luu Nguyen 1987: 11).

been reflected in agricultural land management as well as in the discourses regarding women's issues and peasants' place in society. People in Lang Xanh were part of these discourses; they lived with and engaged with them, in accordance with their conditions and convictions. When male and female farmers in Lang Xanh oriented themselves as men and women, they did so in relation to ideologically informed gender ideals.

Just by walking through the streets of Hanoi, it was possible for me to get some first visual impressions of the 'role models' of femininity most cherished by the state. Large-size billboards with ideological messages pictured women as soldiers, workers, farmers and most commonly, mothers – all of them busily performing their chores for the sake of the nation. Street stands selling journals and newspapers exposed covers of women's magazines that often flaunted models who were incredibly beautiful with white skin, flowing silky hair and a soft smile. These women represented another ideal femininity. They were aiming to please, with their low-key expressions and graceful, yet seductive looks. These pictures held connotations of sexuality, docility and consumerism, rather than struggle, hard work and heroism. At formal occasions, such as a celebration of the 8th of March International Women's day, or a simple meeting in the People's Committee in the commune of Lang Xanh, it was possible to catch glimpses of a whole range of female ideals. This could be seen in the references that spanned from half mythological female warriors, via women soldiers who fought off the US enemy while keeping the rice production intact, to present day exemplary mothers raising well-educated children. Quite soon, I could detect at least three normative types of women, each of them brought into use in accordance with the requirements of the occasion: the fighting woman, the hard-working woman and the family-oriented mother/wife. All three of them shared the spirit of self-sacrifice as they first and foremost fulfilled their duties for the sake of their families and for society at large, and they did it with appropriate grace and dignity.

There is no doubt that Vietnam is a country where officially proclaimed ideologies play a distinct role and people are firmly embodied in these kinds of discourses. As in many revolutionary socialist or communist states, Vietnamese women were – and sometimes still are – categorised as oppressed and exploited, mainly by patriarchy and capitalism. Even after national freedom and the war against America were won, the state continues to surround women with ideologically based movements and campaigns. From this perspective, Vietnamese women have been made into transformable subjects and symbols for national ambitions of modernity, progress, and identity (Pettus 2003: 6–10). These ways of using, defining, and redefining femininity have lead to contradictory and opposing feminine ideals.

For the purpose of a closer understanding of the male and female farmers in Lang Xanh and how gendered identities are shaped in relations to farmland, it is important to go back a bit in time and take a closer look at some important stages in Vietnamese history. It is of relevance to consider the ideological contexts in which mainly 'women' and 'peasants' have been characterised, and in which the ideological meanings of farm work and farmland play significant roles.

Women in revolution and political ideology

During the early 20th century, parts of the Vietnamese population – initially urban intellectuals – had worked up noticeable discontent with the French colonial regime that had gradually taken control over Vietnam since the mid 19th century. The aspiration for national independence went hand in hand with societal modernisation. The general idea was that if any changes were going to take place, both traditional Confucian and more recent colonial ideas had to be fought, especially since they interlaced relatively smoothly (e.g. Bergman 1975, Marr 1981). And when resistance against colonial rule grew more intense and became increasingly organised, ideas about the desired roles and capacities of women were activated – not least their potential to contribute to the opposition against the French colonial regime.

The Confucian heritage

In the first decades of the 20th century, the battle for freedom also meant that the radical Vietnamese of the younger generation opposed the restricting Confucian family structures in their strive for a modernised society (Hue-Tam Ho Tai 1992: 89–91). According to Confucian doctrine (introduced by the Chinese in 111 B.C.) a strict hierarchy regulated family, society and state. Factors such as title, age and gender positioned each individual in a hierarchical position vis-à-vis other persons in every situation. Each position came with well-defined rights and responsibilities and gender traits were seen as intrinsic to each sex. Womanhood – generally defined as inferior to manhood – gave women no other choice than to occupy the lower ranks of society, anything else would be regarded as purely unnatural. The following Confucian-inspired proverb aptly captures this by stating that: '*A woman who meddles in the affairs of court or palace appears as abnormal as a chicken that crows*' (cited in Bergman 1975: 25).

However, in certain situations, by way of age or title, women could be superior to men, though generally, Confucianism put the vast majority of women under the rule of a superior male subject. This relationship was explicitly formulated in what is usually referred to as the *three submissions*. This meant that a daughter had to obey her father, a wife should follow her husband, and an elderly woman, or a widow, was expected to conform to the decisions of her son (Marr 1981: 192). Throughout these all-embracing stages of a woman's life, a myriad of rules circumscribed her behaviour and set the standard in every detail. They were often stated in the form of epigrams and with the purpose to breed women into good daughters, wives, and mothers (Hue-Tam Ho Tai 1992: 93). In essence, these rules boiled down to the women's *four virtues*: 1) labour/housewifery, 2) physical appearance, 3) appropriate speech, and 4) proper behaviour (Marr 1981: 192). Within these clearly demarcated and family oriented positions, women were supposed to live, act and work – ideally within the confinement of the family house.

Since Confucian gender ideals were still setting the standards in the earlier part of the 20th century, the manners of rural women – who took more frequent part in activities outside their homes and worked closer to their husbands and other men – had, so far, won little appreciation from the urban quarters (Bergman 1975: 21–25; Hue-Tam Ho Tai 1992: 94; Marr 1981: 197). Despite the relatively loose grip of Confucianism over rural women, it was men who generally had the privilege to make major decisions and to control the assets of the household. This commonly included controlling the division of labour and any land issues within a farm household. Property rights, inheritance, and order of family succession were, only with some exceptions, unfavourable for women (Nhung Tuyet Tran 2008).

In general, any behaviour reminiscent of self-interest was discouraged by Confucian doctrine. The well-being of the family and of society should be top priority at all times. Gradually, the radicals of the budding colonial resistance considered that such constraining norms gave little freedom for the individual to make choices in life and to alter the 'inevitable' responsibilities attached to one's gender and position[2].

'Women's issues' and colonial resistance

At the time leading up to the 1945 revolution, the issue of women's roles in the family and in society was used in a wider sense to discuss cultural and political questions. Under the cover of debating women's role in society, more sensitive political and anti-colonial issues could be discussed by opponents of French colonial rule (Hue-Tam Ho Tai 1992: 88–89). According to the historian Hue-Tam Ho Tai, the beginning of the Revolution in Vietnam was really emerging from a discontent with the structures of the traditional family. The younger radicals who questioned the Confucian legacy of gender and family relations were in effect expressing doubt about the entire societal order, as they saw that equivalent power structures were governing family and society. French colonial rule, as well as traditional Vietnamese views on family and society, was often expressed in patriarchal terms. The position of women as victims of oppression thus appeared as a useful vantage point for opposition. Women, in their relatively repressed positions, came to symbolise the tension between generations, between the masses and the rulers, and between the colonial power and the conquered country (ibid.). Consequently, when the Indochinese Communist Party (ICP) was founded, in 1930, the struggle for gender equality was included in the main revolutionary goals (Marr 1981: 235–37).

The colonial regime did what they could to repress the resistance; in the late 1930s, many men and women were imprisoned, tortured, and executed. Some of the women who continued to be involved in the resistance movements, and who had managed to avoid captivity, continued their fight against French rule as armed soldiers (Marr 1981: 244–46).

Despite the risks that many women were willing to take in order to fight for independence and a new order of society, in practice, male Party members did not hesitate to let women perform what was still considered

2 See Hue-Tam Ho Tai, *Radicalism and the Origins of the Vietnamese Revolution*, Harvard University Press, 1992, for an elaboration on this subject.

to be female duties. Women were encouraged to give up their traditional family roles to join the revolutionary forces, but the revolutionary woman was given no place in the existing family and gender relations. In real life, it was difficult to be good daughter or wife in combination with revolutionary activities, and there was no corresponding debate urging men to fill in for the women and to share their domestic duties. Those women who actually gave up their old lives and became revolutionaries often had to remain unmarried (Hue-Tam Ho Tai 1992: 211–12).

The Vietnam Women's Union

The same year as the ICP was formed, Nguyen Thi Minh Khai, a female party member, established Vietnam Women's Union. This was the first national women's organisation of the country (Marr 1981: 243). The Vietnam Women's Union is still an integral part of society with its national structure, which starts at the head quarters in Hanoi and is distributed down through the provinces, districts, and communes, with branches in more or less every village. The structure of the organisation resembles the present day Communist Party,[3] with cadres (*can bo*)[4] at each level passing information and decrees to the lower levels and mediating reports on results and activities in the other direction. Despite the fact that the Women's Union is not formally an organ of the government but a mass organisation, the purpose and intention of the Union and the government have always largely coincided. In practice, the Women's Union of today is a body for promoting women's concerns within the government (Tran Thi Van Anh and Kaime-Atterhög 2000: 4). Through their work and their publications, the Women's Union have

3 Vietnam's Communist Party has its origins in the Indochinese Communist Party (ICP), founded in 1929, which (publicly at least) was dissolved for some time during the war with France in order to encourage non-communists to join the nationalist Viet Minh cause. Re-established in 1951 as the Workers Party of Vietnam, in 1976 following the reunification of the country it was renamed the Communist Party of Vietnam.

4 'Cadres are Vietnamese citizens who are elected, approved or appointed to hold posts or titles for a certain term of office in Party or state agencies or socio-political organizations at the central, provincial or district level, and a number of types of commune-level cadres who are on the state payroll and salaried by the state budget.' As defined by the Law on Public Employees, effected on 1 January 2012 (http://vietnamlawmagazine.vn/news/institutions-on-civil-servants-in-vietnam/1359bb1e-41f6-4a28-9364-a2e7a33e5c3c.html).

tried to supply women at all levels with social and moral guidelines and a will to strive to become good citizens in society (Pettus 2003: 29). In 1984, Arlene Eisen wrote enthusiastically about the work of the Women's Union. She reasoned that women, especially peasants, who had been enmeshed in patriarchal structures for ages needed to have their eyes opened before they could fight for the revolutionary causes of national and female liberation, and that this could be achieved through union activities (Eisen 1984: 119). Some of the enthusiasm seems to have worn off, not least due to the slow progress of female emancipation in Vietnamese society. Werner's view of the Women's Union from 2002, for example, gives voice to a more critical stance of the Union's work. She argues that under the pretext of looking after women's best interest, the main objective of the Union still concerns the state rather than women's interests. Moreover, women's concerns are singled out as separate issues instead of forming a part of all societal matters. The model of gender relations promoted by the Women's Union is 'essentially paternalistic and instrumental' (Werner 2002: 44).

Making good socialist women

Instead of losing their significance, numerous Confucian and traditionally instigated feminine virtues were extended by the communist ideology to include the hard work, devotion, and proper behaviour that were needed for the cause of the revolution and for the building the nation. In the aftermath of the 1945 revolution, when the Democratic Republic of Vietnam was founded, and after 1954 when the French finally were defeated, the situation of women was becoming quite ambiguous; the plight of a contradictory double burden started to emerge. Women were supposed to learn to manage politics and state affairs as the equals of men, but in practice, they were not relieved of any of their household, family, and farming duties. Nevertheless, the Vietnamese communists kept arguing that in a socialist structure, women should no longer be exploited or be deprived of power. To some extent, this proved to be true when a number of women did gain significant control, primarily in the lower levels of the state, Party, and official organs. Women were also permitted equal legal rights, and their protection within marriage was enhanced. In late 1959, the communist regime passed the Law of Marriage and Family. This law banned such things as polygamy, wife-beating and arranged marriages, and endorsed men's and women's equal rights to for instance divorce, custody of children, and property (Malarney 2002: 149; Pham Van Bich 1999: 58). Working conditions were improved, by for example introducing maternity leave, at least on paper, and women received education to a much greater extent. In certain fields of work, like teaching and medicine, women could now become professionals (Marr 1981: 240, 242, 249).

White has pointed out that it was no coincidence that the changes in marital law carried out in the late 1950s concurred with the Party's urge to change from the dominating household farming into a collectivisation of land and farm labour (White 1988: 169). The traditional family structure had tied the peasant families tightly together as a social unity wherein the needs and interests of the members were controlled. The family also provided an important base for the peasants' household economy. Access to land, labour, and means of production and sustenance were managed within the household and extended family unit. Up until now, as mentioned above, labour division and decision-making had been male biased, and marriage was an important way for a household to recruit female farm labour in the form of daughters-in-law. Traditional practices like these – with women generally performing much labour but exercising little control – made more women than men positive towards the formation of agricultural collectives. As 'mangers' of household farming and heads of households, men could dread a loss in their privileges (ibid.).

With a collective work force, some women saw a chance of having their farm work reduced by a more equal division of labour among the members, and gain more recognition for their labour. For some men, this would mean an increase of their farm labouring efforts, which had traditionally mainly comprised ploughing and preparing the fields. In areas with a strong Women's Union, the advantages of jointly used resources as a way of improving life for both male and female peasants, was brought forth in order to convince the rural population of the advantages of collective farming (Kerkvliet 2005: 75). Men, as heads of households with extensive decision power, were sometimes difficult to convince when authorities used arguments in favour of gender equality (White 1988: 169).

The message the communist regime wanted to convey to women was that the sacrifices and contributions they made to the nation were one and the same as – or even a prerequisite for – carrying out their family responsibilities; i.e. in order to be a good mother or daughter-in-law, a woman had to labour for the collective needs and *vice versa* (Pettus 2003: 28–29, 32–37). In this manner 'the party aimed to wed young women's hearts, minds and bodies to the socialist cause' (ibid.: 39).

In 1961, during the early years of the Democratic Republic of Vietnam, the Women's Union introduced the *five goods* movement, tracing back to the Confucian *four virtues*. The core message to women was that they could and should 'do all things, and do all things well' (ibid.). More exactly, the five points spurred women to 1) fulfil the goals of production and economise well; 2) follow all state policies and laws; 3) participate in management; 4) advance in their studies; 5) raise their families and educate their children well (ibid.). The resemblance to the normative lists of Confucianism is rather striking and the expectations of the state on women seemed to have increased.

Even so, Wiegersma argues that any form of equalisation of power in the rural villages mainly took place among men, or male-headed households, because men were the main recipients in movements and campaigns concerning training within collective production and management and, compared to pre-revolutionary rural society, their income became more equally distributed through the farming collectives (Wiegersma 1988: 155). Even if some women made advancements in the local power structure, they very seldom reached the higher positions (Kleinen 1999: 125). It was, however, possible to see some traces of the officially supported gender equality. In theory, men and women could join the collectives on equal standards. Even if women's equality was not always reflected in reality, women had now made some advances compared to the villages of colonial and pre-colonial times, when women were not allowed in the village meetings and communal land was given to male household heads only.

Women at war again

In 1965, USA launched the first serious air strikes against the Democratic Republic of Vietnam. The Americans had been present in South Vietnam ever since France was defeated in 1954 and they supported the anti-communist regime controlling the southern region. In order to mobilise the people of the North, the Communist Party intensified their call on the high morals of the citizens and their willingness to sacrifice themselves for their country.

All the campaigning aimed at the Vietnamese women during the war against USA included the underlying idea of making sure that they were doing all they could to contribute their labour and fighting spirit when the nation needed it. Just like when France was fought, the female combatants could, at least temporarily, abandon some of the traditional female traits, like docility and family obedience. Women who adjusted to the severe situation and made personal sacrifices were held in high regard – and sometimes rewarded by Ho Chi Minh himself – as role models for other women to follow (Ngo Thi Ngan Binh 2004: 52). Both men and women were expected to sacrifice their personal lives for the love of the country. Exhorted by the Party and the state, men were compelled to show their solidarity by following the '*Three Readinesses*': i.e. they should be ready to fight, to join the armed forces, and to take up whatever task the country thought necessary (Luong 1992: 202). Aimed specifically at rural women, the regime simultaneously issued a proclamation of the '*Three Capacities*'. These involved the capacity to take care of the family while encouraging men to fight; to take charge of every duty and task while the men were away; and finally, to serve the war and be prepared to fight themselves (ibid.). Women should also promise the departing men their fidelity. The Confucian heritage is made visible again in the epigrammatic form used to convey ethical messages – but also in the actual content, urging the citizens to be loyal to the nation and rise above their personal needs in gender appropriate ways (c.f. Malarney 1997: 907–08).

Keeping up the work in the farming collectives was the most important contribution to the war effort made by women in the countryside. As an effect, the war had the greatest impact on labour division in the collectives (Wiegersma 1988: 170). The fact that so many of the male villagers and collective members were absent during the war made it possible for the remaining women to hold positions earlier occupied only by men (ibid.: 160). With the men away at the front, women simply had to undertake many of the duties traditionally performed by men. They ploughed and applied manure, and they took on the role of managers of the agricultural collectives when needed. Yet, as we will come back to, the impact was only of momentary value (ibid.: 162–63).

A rise in the ideological position of rural women

As the war against USA continued, there was a shift in the propaganda towards the female population in the early 1970s. Instead of encouraging women to adopt a particular moral behaviour, criticism of certain conducts by women emerged. Focus was aimed at urban women developing a taste for the 'good life'. Stimulated by influences from abroad, the shift toward the liberation of women had, according to the Party, led to younger women's desire for leisure time, nice clothes and personal beauty. In the eyes of the regime, these young women were slipping out of the moral spheres of family and good citizenship and they did not contribute to any production or collective labour (Pettus 2003: 54–55). This kind of frivolous behaviour was not equality, the Party officials stressed. These women 'mistook the fruits of Revolutionary liberation for personal freedoms – an error that jeopardized the values of domestic femininity' (ibid.: 59). Thus, again there was a clear linkage to Confucian directions of proper female conduct. In order to counterbalance the growing 'misunderstanding' of gender equality, the Party and the Women's Union carefully pointed out that it was the duty of women to safeguard the fine qualities of Vietnamese tradition (ibid.: 59, 61).

Judging from the Party line of reasoning of this period, rural women stood a better chance of coming near the ideal female model than urban women. Family relations in the countryside were less influenced by more advanced ideas of female liberation and most of the women spent their days working in the fields of the collective farm to support their families. Rural women were still viewed as lacking in 'refinement' and 'modernity' and being in need of education to become 'progressive', but they were on the right side of the moral line as far as the Party was concerned (ibid: 59). The 'feudal' and 'backward' ways, bearing clear traces of Confucian patriarchy, which rural women sometimes were accused of supporting, were no longer exclusively declared as incorrect. Instead, the less radical behaviour of countryside women provided a counterbalance to certain kinds of female conduct, which was threatening to slip out of control (ibid.: 60).

The 1970s can be said to be the ideological heydays of female peasants. They worked hard, either in the rice fields or in the battlefields, and they struggled to combine this with caring for children and other family members at the home front.

Economic renovation and a new kind of woman

The 1980s was a time of recuperation from the damages of war and a time when the building up of a prosperous society was to get going. As a result, the Vietnamese state shifted its activities, from trying to drum up people's motivation to fight against enemy assault, into making efforts aimed at economic restoration. In order to meet the pressing demands for increased productivity, not least of agricultural products, the state again started to modify its standpoint on the role of women.

Factors such as failing to assimilate the market economy of the southern part of the country, loss of economic aid from the US (in the south) and from China (in the north) and falling outputs from all sectors of the state combined with decreased loyalties towards centrally made decisions made the need for economic changes unmistakeable (Dang Phong and Beresford 1998: 103–104). In December 1986, at the sixth Party Congress, it was decided that Vietnam was going to abandon the Marxist planned economy for the benefit of a 'multi-sector commodity economy' where the private sector and foreign investments should be recognised as legitimate parts of the economy (Ljunggren 1997: 15). *Doi moi*, which literally means renovation, was introduced. The goal was to renovate the existing system, not only the economy, without threatening the position of the Communist Party (ibid.: 12). The insufficiency of the agricultural collectives led to an ideologically modified reinstatement of the market-oriented economy, where the household as a productive and reproductive unit regained its central position.

Gradually, *Doi moi* started to reinstate Vietnam into the global economy and new standards, demands, and desires appeared among the citizens. The government embarked on a tricky path where the balance between economic progress and the maintenance of national culture and morality had to be carefully managed. The regime wanted to promote household improvements in line with global economic and social standards while, at the same time, trying to control and define national morality and culture to avoid 'immoralities' and 'social evils' connected with a market economy (Pettus 2003: 80, 88). An example of this delicate balancing was displayed in the work of the Women's Union in this renovation era. In order to enhance the level of modernity, they defined politically instigated spiritual as well as material goals and standards for women to try to achieve. To develop 'prosperous families', capable of caring for economic as well as moral needs of its members, became an explicit state concern (Bélanger and Barbieri 2009: 13–14). The family – which is still mainly the responsibility of women – was their main target because this was where 'the problems of economic fitness and national moral health come together' (Pettus 2003: 80–81).

Again, Vietnamese women are encouraged to be bold, progressive, and yearning for more knowledge in order to contribute to national progress. By adopting the state's ideas of desired female capacities, women are expected to transform themselves into modern Vietnamese women. But aspirations of 'progress' and 'modernity' can be difficult to balance, as they are sometimes inconsistent with other, more family-oriented, female ideals. This is summed up rather well in an interview with the Chairperson of the Women's Union in Ho Chi Minh City, published in *Images of the Vietnamese Woman in the New Millennium* (Le Thi Nham Tuyet [ed.] 2002). The Women's Union leader, Nguyen Thi Lap, is asked to comment on the relevance of the *four virtues* (proper labour, appearance, speech, and behaviour) for women of today. She believes that the moral standards expressed in the *four virtues* are still valid today, even though they should be applied with today's society in mind. In light of today's consumer society, she expresses her worries about young girls losing their dignity and self-control when the attraction of material wealth becomes too strong, and in the worst case, they might fall into prostitution. She stresses that desire for wealth or career should not obstruct women from their most important task, which involves being a mother and a wife. Women should primarily be knowledgeable in organising their families and in understanding the workings of marriage. Altruism, she concludes, is an important trait that will enable women to put the community before self-interest (ibid.: 7–8).

Rural women in need of change

An ideal picture of women promoted by the Vietnamese Women's Union implies that Vietnamese women are not anymore encouraged to be hardworking peasants, producing for the benefit of the nation. Women are expected to operate largely within the family and household sphere where they provide care, moral guidance and support, preferably in combination with qualified work, i.e. work which requires an education. These ideals are transmitted to peasant women and they have to relate to them in one way or another (see e.g. Pettus 2003; Drummond 2004). Working at home is no longer denounced as a source of patriarchal oppression of women. Women are increasingly expected to combine their double responsibilities, for production as well as for reproduction, within the domain of the household (Truong Thanh-Dam 1997: 83–84). This development is in line with the revival of Confucian ideology, amalgamated into communist ideology, placing women firmly within the household and in charge of the social, moral, and emotional values of the family (e.g. Ngo Thi Ngan Binh 2004: 47–73).

The position of rural women is ambiguously linked to the female ideal of present day Vietnam. On the one hand, they perform most of their labour within the realm of the household and devote much of their time to nourish and care for family members. Furthermore, as previously indicated, they have claimed comparatively little of the unsolicited kind of female liberation that urban women might ask for. On the other hand, peasant women are not perceived as the kind of well-educated, yet delicately mannered homemakers that the regime has been arguing for (e.g. Drummond 2004: 158–78). The ideological restoration of the mother and housewife into the realm of the family has taken place with the urban, educated, consumer family as the role model. Judging from some of the Vietnamese publications[5] dealing with the issue of rural women in the last 15 to 20 years, peasant women are in urgent need of adjusting to new circumstances. It should also be added that it is not easy for rural men to fit into the role of the successful breadwinner, who brings home a substantial cash income and leaves a part of it to his wife so she can use it wisely for the benefit of family harmony and prosperity.

The changing roles of 'peasants' and land

Along with 'women', the category of 'peasants' was at the centre of attention for the communist leaders during and after the battles for national liberty, as well as during the formation and progress of the socialist state. The support of the enormous group of peasants was necessary and like 'women', 'peasants' – and farmland – have continued to be defined and redefined in order to fit the requirements of the times.

Building up to revolution: trying to radicalise the peasantry

When the radical male literati first began debating the desired morals and customs for a Vietnam free from colonial rule, the countryside

5 E.g. Do Thi Binh 1997: 7–9; To Duy Hop 1997: 15–19; Tran Thi Van Anh and Le Ngoc Hung 1997: 131–52; Hoang Ba Thinh 1998: 4–27; Nguyen KimThuy 2002: 41–43; Le Thi 2003: 109–16.

was commonly pictured as in great need of refinement and enlightenment (Hue-Tam Ho Tai 1992: 49, 104). This had its roots in the still influential Confucian idea of an educated elite male leadership giving moral guidance to their subjects. Ever since Vietnam became subject to Confucianism, starting in 111 BC with the conquering by the Han Dynasty, there had been a marked disparity between the peasants and the elite, probably accentuated by the peasants' lesser observance of Confucian manners (Duiker 1995: 167; Moise 1983: 147). This tradition characterised the relation between the elite and the many peasants of the country (Marr 1981: 28).

In the anthropologist Hy Van Luong's work on the revolution and transformation of the northern village Son-Duong, a son of a rural elite family describes how his father kept him and his brothers from engaging in manual labour. Because they were being educated to become members of the respected literati they were urged not to cross the distinct line that separated physical labour from mental efforts. The same restrictions did not apply to his female relatives, who regularly participated in farm work on the land belonging to the family (Luong 1992: 68, 2010: 69–71). Village power was predominantly executed by men of the elite, who performed little or no manual labour. During colonial time, the French actually reinforced the power of the patrilineal-based male elite in the villages when they replaced the former councils of notables – who used to be chosen by village men to decide over village matters and communal land – with male representatives from elite families chosen by the colonial rulers (Bergman 1975: 40, Luong 1992: 71, 78, Wiegersma 1988: 53).

The hierarchical and tradition-based village structures, which the French regime had maintained, were targets for the anticolonial resistance forces. According to them, peasants needed to develop a critical stance toward the old order of power and loyalties. For example, what was perceived as the preoccupation of peasants with family affairs and their comparatively huge expenditure on rituals, such as weddings and funerals, attracted criticism from the revolutionaries (Pham Cao Duong 1985: 51, Pham Van Bich 1999: 79). However, from a peasant perspective, the reciprocal family rituals and relations were important for gaining support in future situations in general and more significantly as an insurance in case of upcoming hardships. The revolutionaries, on the other hand found the modest extra-family and extra-village perspective of the peasants as a sign of lacking knowledge and revolutionary spirit (Pham Van Bich 1999: 81). During the mid 1920s, there was an internal debate in the Party, discussing the revolutionary qualities of the peasantry, which from a Marxist-Leninist point of view was rather limited (Luong 1992: 85–86, Ngoc-Luu Nguyen 1987: 16). According to this vein of communism, which was where the Vietnamese found their inspiration, the peasants were an unreliable class because of their property-based, self-sustaining mode of production and their alleged inclination to preserving old customs and social hierarchy. Nevertheless, the Vietnamese communists understood the potential of the peasants. The conclusion was that peasants needed to be educated, politically organised, and made aware of class issues (Brocheux 1980: 77–79; Ngoc-Luu Nguyen 1987: 16; 42–43, Wiegersma 1988: 92).

In 1937–38, Vo Nguyen Giap and Truong Chinh, two eventually reputed communist leaders, addressed these issues in the publication 'The Peasant Question'[6]. They worried about the reluctance of the peasants to see their situation through the lens of class analysis. According to them, poor and landless peasants had a too accepting perspective on their situation and tended to blame their bad fate rather than social injustice for having to work on someone else's land for a very meagre compensation from the landowner. Some peasants even expressed a certain amount of gratitude for the opportunity to rent land from some rich neighbour or relative[7]. 'Unlucky fates' were also commonly handled

6 Truong Chinh and Vo Nguyen Giap. *The Peasant Question*. Ithaca. N.Y.: Cornell Southeast Asia Program, 1974. Translated by Christine P. White of *Van De Dan Cay* (Hanoi: Su That, 2nd ed., 1959). Originally published in two volumes under the pseudonyms Qua Ninh and Van Dinh (Hanoi: Dan Chung, 1937 and 1938).

7 Despite the peasants' assumed lack of revolutionary features, smaller groups of peasants had instigated uprisings against local French, or colonial-friendly, authorities. Peasant revolts were actually a tradition dating far back in Vietnamese history, when displeased peasants had carried out violent uprisings in order to improve hard conditions (Wiegersma 1988: 22, 42, 87; To Lan 1993: 186). In 1930, the renowned Nghe-Tinh uprising took place when large numbers of peasants, both men and women, carried out demonstrations, created their own administration with peasant-led boards to manage land that was taken from the local landlords (Wiegersma 1988: 95; Bergman 1975: 52). 12 months later, by the use of force, the French resumed their command over the two provinces (Nghe An and Ha Tinh) while placing the blame for the unrest on the ICP (Wiegersma 1988: 95–96). These

with the help of worship and prayer to various deities. The revolutionaries referred to this as superstition and a substitute for societal awareness. Therefore, alongside traditional village and family hierarchy, peasants' spiritual matters became targets of communist campaigning (Pham Van Bich 1999: 79, White 1983: 75).

Revaluating the peasants

One of the major steps in creating a new socialist society was to alter the control of land and the old power structures in which landowning men held influential positions. Under the influence of Confucian doctrine, the Vietnamese villages had been turned into stronger and more centralised units, where local patriarchs expanded their authority. Each village had a council of notables – where women were not allowed as members – with the power to decide over village matters. Communal landholdings constituted the economic base for the village. Some of this land was set aside to provide for the costs of administration, schooling, cultural activities and support for the needy. The rest of the communal land was regularly distributed to adult men in the village, usually for a period of a few years at a time. This land was subject to tax, payable to the village administration (White 1988: 168). Despite varying levels of equality in allocation, communal land established a basic platform of security among the households (Nguyen Duc Nghinh 1993: 210).

The system of communally owned land had been supported by indigenous governments in order to secure tax revenues and a certain level of welfare for the population. It also implied that villagers could not sell all the land that they cultivated and that the mobility of those entitled to this type of land was limited (Moise 1983: 148). The amount of communal land in a village could vary from only c. 15 *sao*, to all the land at a village's disposal. Each village also had communal land which revenue was used to cover the expenses of common religious rites (Nguyen Duc Nghinh 1993: 207–08).

After a shift from one dynasty to another, larger landholdings were often confiscated from landlord families and ended up in the hands of other influential families or as village communal land (Wiegersma 1988:

events contributed to bring the communists and the peasants closer together (Marr 1981: 378; Moise 1983: 152; Wiegersma 1988: 95–96).

50). Such a cyclical landowning pattern did have some advantages for the peasants because it prevented landlords to acquire great landholdings for very long periods of time, and even the least fortunate could receive a field for rice cultivation from the share of communal land (Pham Cao Duong 1985: 24–25). This pattern continued all the way until the 1945 August Revolution, which put an end to any land accumulation by individuals or families (To Lan 1993: 160).

Most likely, village communal land contributed greatly to the high level of autonomy of the rural northern Vietnamese villages. Duiker argues that 'Chinese cultural veneer' (Duiker 1995: 167) was rather thinly painted on the Vietnamese villages, and that it is here that we can see the beginning of what was to become a constant characteristic feature of northern Vietnamese society, namely the dichotomy between a small and influential elite inspired by Confucianism, sided later on by French administration, and the rather more communal and egalitarian traditions of the 'commoners' in the villages. This ideological gap, Duiker points out, has been used by Marxist historians to argue that the feudalists oppressing the masses were foreign, or influenced by foreign rulers, as a contrast to the ordinary villagers and peasants, who were assumed to embody more 'pure' Vietnamese traits (ibid.).

Before the French colonial rule, the land holdings of villages in northern Vietnam consisted, apart from communal land, of state and privately owned land. Even if not all the village dignitaries owned substantial portions of land, they had access to communal land, and those who did own much land, i.e. landlords, were closely connected with the village administration (Nguyen Duc Nghinh 1993: 219, 221). So, there was a coherence between power and land, and with the French colonial rule, the amount of privately owned land increased. The French tapped into the already existing tax administration where the villagers had paid tax on their land to the local administration, making all villagers, regardless of their level of wealth, pay taxes. They also created a land market where small-scale farmers stood little chance to obtain any newly cleared land (Wiegersma 1988: 73–74)[8]. Just before the 1945 revolution in northern

8 In 1931, 19 per cent of arable land in colonial Vietnam (Tonkin, Annam and Cochinchina) was owned by French concessions. The highest concentration of concessions were found in Cochinchina (the south), where 80 per cent of the peasants were tenants, either of French or Vietnamese concessionaries, by the end

Vietnam, the French citizens, the Catholic Church, landlords, rich and middle peasants[9] owned c. 63 per cent of the land in northern Vietnam (Moise 1983: 164). Luong reports from Son-Dong village in the Red River delta of North Vietnam that the colonial period was characterised by a class-structure with significant gaps between the status and wealth of the households. In 1954, before land reforms had been implemented in the village, politically perceptive landowners had already disposed of some of their landholdings. At this point in time, 2.6 per cent of the households owned 23 per cent of the rice fields, while landless and poor peasants made up 70 per cent of the households. This situation resembled other northern villages at that time (Luong 2010: 62).

The land reformation campaign that the communist regime carried out in 1953–56 was as much a social and political scheme as it was an economic measure. The idea was to attain 'real class struggle to weaken feudalism economically and smash it politically.' (Moise 1983: 170). Mass mobilisation cadres were sent out to the villages to make contact with poor peasants and, through their knowledge of local circumstances, classify the villagers according to class demarcations based mainly on landholdings, economic status and whether they practiced usury or exploited labourers and also note their attitude to the French and to the new republic and the resistance against the colonisers. The rural population in the northern villages were divided into labourers/landless peasants, poor peasants, middle peasants, rich peasants, and landlords. Labourers and poor peasants constituted 50–60 per cent of the rural population. They had little or no land and were subject to usury, unfavourable land rentals or sharecropping, and exploitative labour conditions. They were the main recipients of land. The landlords were singled out as the

of French colonial rule (Wiegresma 1988: 71–75). In the province of Phu-Tho, were Lang Xanh is situated, c. 22 per cent of the arable land was controlled by French settlers or Vietnamese collaborators in the 1930s. Most of this land comprised tea plantations (Luong 1992: 43–44).

9 Landholdings and the degree of land-related exploitation were the two decisive criteria for classification of households. Landlords had large amounts of land and engaged in exploitation through e.g. usury, sharecropping and use of hired labour. Rich peasants had sizable land but mainly used household labour, even if they also hired labour, practised usury, etc. Middle peasants first and foremost used their own labour and had sufficient access to land, even if they occasionally hired labour or rented land from others (Luong 2010: 167–68).

main enemies by the communists. They had made their living by renting out land, or they had hired labourers to work on their land at poor wages, often without contributing any labour themselves. Many of the landlords had also been collaborating with the French colonial regime (ibid.: 210–13). The idea of the land reform was that all land and other property belonging to landlords, Frenchmen and others loyal to them, would be confiscated and distributed to landless and poor peasants. Rich peasants would have some of their land redistributed and middle peasants could lose or receive land, depending on how much land they had in relation to the other villagers (ibid.: 211–12).

Through the land reform, the old village structures were drastically changed. The power hierarchy was virtually turned upside down when the former landowners and patrilineal members of the village councils were deprived of their power and assets. From the communist perspective, the old elite had nothing to contribute to society; the model citizen was a manual labourer (Malarney 2002: 28). Just like in the case with rural women, to raise the status of the peasants, physical labour was idealised as the foundation of a new society. However, it was important that such labouring occurred for the sake of national progress, not for personal revenues. All of a sudden, the poorest peasants were appointed by the reform cadres to hold many of the leading positions in the villages. Poor peasants and workers were claimed to be the only ones to be trusted as true friends of the communist regime (Moise 1983: 198, 210).

The strong ideological conviction and the frantic pace in which the land reform was implemented resulted in incorrect classifications and excessively cruel behaviour – even killings – towards some people classified as landlords. The occurrence of unfounded or exaggerated accusations of landlordism resulted in the 'rectification of errors' campaign in 1956 (ibid.: 213, 237). During this campaign, some of the wrongly accused landlords (*dia chu*) were reclassified and somewhat compensated for their losses. Nevertheless, the old social structure of the villages was more or less completely shattered; the members of the village councils had been replaced with loyal – sometime inexperienced or incompetent – cadres of 'pure' class origin. At the time of land reformation, fear, grief and hate had permeated many villagers and it would last for a long time to come (Kleinen 1999: 105, Luong 1992: 195, Moise 1983: 232,

252). On the other hand, the land reform did achieve many of its ends. Land was redistributed and many poor and landless peasants received land, landlords were dispossessed of their land, and much of the village authority passed into the hands of peasant cadres. Many formerly poor peasants had their circumstances improved and the support for the Communist Party among the masses of peasants grew stronger. But the costs for getting there had been high in terms of broken relations, incorrect accusations, ruined lives, deaths and (sometimes) unskilled leadership (Kleinen 1999: 104; Luong 1992: 195–96; Moise 1983: 232).

Making peasants collective

All along, the long-term goal of the Communist Party had been to turn peasants into workers, i.e. proletarians, on land that was not privately owned, but instead organised in agricultural collectives. Ideological as well as practical reasons were behind the decision to collectivise the agricultural sector in the end of the 1950s. The aim of the communist regime forming an advanced, industrialised, state – initially financed with state-controlled agricultural surplus – was built on ideas based on non-capitalist methods and economy. This intention coincided with the regime's need to control food supply, and the possibility of mobilisation troops to fight in the south (Kleinen 1999: 109, Wiegersma 1988: 146). The Communist Party thought that agricultural collectivisation would bring about efficient use of labour and other means of production, which would lead to an industrialised and modern agriculture with an output much greater than the small plots of land, farmed mainly by household labour, could ever produce (Brocheux 1980: 77–79; Ngoc-Luu Nguyen 1987: 16, 42–43). Between 1958 and 1960, 40 000 agricultural collectives were set up in North Vietnam (Kerkvliet 2005: 65). Moreover, the communist regime argued that the countryside ran the risk of falling back into the old power structures if a land market was allowed to work its way freely. This was a reaction to the reoccurrence of poor peasants selling all, or parts of, their land to richer peasants when the crops failed. In order to persuade poor peasants to give up the little land they had received to the agricultural collective in the preceding reformation campaign, they were ensured by the communist leadership that a bad crop or other misfortunes would not ruin them of their means of support. Instead, the agricultural collective would prevent the risks that could threaten to return them to landlessness, poverty, and debts (Kerkvliet 2005: 74; Wiegersma 1988: 146–47).

Furthermore, with the end of small-scale household farming, the labour and production of the peasants (not least the female) were assumed to be liberated from the confinement and control of the patriarchal family and household and thus, could be directed to advance the entire society. It was thought that the persistently family oriented peasants would be more collectively oriented if they were joined in collectives. Labour exchange or other communal activities, like festivals and other village ceremonies, were not unfamiliar to the rural population (see also chapter 5). Still, many peasants were rather hesitant toward farming collectives, mainly because more or less the entire agricultural production, including all the fields, the tools and other production means had to be handed over to the agricultural collective who would manage the farming activities (Kerkvliet 2005: 9–12; Wiegersma 1988: 145–46). The reason that farmers joined the collectives was largely a question of succumbing to the pressure of local authorities to follow the policy of the state, combined with faith in the politics of the Party, which so far had seen to their needs (Kerkvliet 2005: 73).

Wartime and intensified collectivisation

In Kerkvliet's work on the collectivisation process it is evident that the efforts that the peasants made during the time of collectivisation would not have taken place if the country had not been at war with USA (c. 1965–75). Farming cooperatives were central units for recruitment of soldiers and for feeding the troops at the fronts (Kerkvliet 2005: 86). Krekvliet describes how the labourers engaged not only in saving the country from the attacks of United States, but also how they worked to support their friends and relatives who were directly involved in the war. He reports that in Phu Tho province (where Lang Xanh is situated), 60 per cent of the families had at least one relative actively participating in the war effort (ibid.). This was an incentive for the villagers in the cooperatives to persist in their work. 'Meanwhile, for the people directly involved in the war effort, the cooperatives were assurance that their families back home would have something to eat precisely because produce was divided not just according to one's work but also according to need.' (ibid.). Considering the fact that more than 80 per cent of the population were peasants (ibid.: 8), it is understandable why the accomplishments of rural men and women in warfare and in farming turned them into important and respectable citizens at the time.

When the war against USA was finally won in 1975, the peasants' interest in collective farming declined and was instead turned to their household garden plots (Loung 2010: 185). The regime, on the contrary, wanted to merge existing collectives into even larger ones with the hope to increase the control of production and divert the peasants' inclination for household and village-based farming (Nguyen Duc Nhuan 1983: 365–66). These steps, taken by the Communist Party, did not reduce the effects of the cooperatives' inefficient land-use and the farmers' decreased willingness to work collectively. Instead, production continued to plummet (Kerkvliet 2005: 149–51, Luong 2010: 185–86). As examples of the ineffectiveness found in many of the farming cooperatives, Loung reports that in the village of Son-Duong, cultivable land was used for public work projects such as warehouses, paddy-drying courtyards and live-stock farming (Luong 2010: 186). Kerkvliet tells about the distrustfulness that appeared when people from larger areas were expected to work together in groups of more than 100 labourers. People started to suspect that other members of the large working brigades only 'went through the motions of farming'. 'When brigade leaders were present, people worked diligently. When not being watched, however, many people slacked off – or were suspected by fellow workers of doing so. Increasingly, people asked themselves, why work carefully when others work carelessly?' (Kerkvliet 2005: 147).

The large labour brigades were divided into units performing highly specialised duties – there were for instance brigades for rice-growing which subdivided into brigades for cultivation techniques, ploughing and harrowing, hydraulics, mechanics, transport, and for transplanting, tending and harvesting. Thus, the wide-ranging tasks of farming became fragmented when the state urged the cooperative to specialise in one or two crops or in animal breeding (Nguyen Duc Nhuan 1983: 367, 368). The enlargement of the collectives also generated a huge body of officers, controllers, timekeepers and so on, which transformed the labourers into passive workers receiving and carrying out orders from above (ibid.: 370). Another problem was the leadership of the cooperatives. Facts or rumours about corruption were not unusual and the cooperative members were vulnerable to the leaders' morality and knowledge, as well as to their ability as managers. As Kervliet concludes, 'careful farming by an individual had a chance of paying off only if brigade and cooperative leaders coordinated labor and other requirements and all or nearly all members in the brigade worked well, not just one day but every day and every step along the way' (Kerkvliet 2005: 111–12).

It is possible to claim that the enlargement of the farming cooperatives in the north influenced the bond between the farmers and the fields in a negative way. As a result of this move, the peasants became largely disengaged from the actual outcome of their work. The detached farming process made the production of the collectives largely unpredictable, and there was little correlation between the work that the cooperative farmers invested (or avoided to invest) on the fields, and having rice to eat in the bowls – too much depended on factors that each farmer could not control (Nguyen Sinh Cuc 1995: 78, 80). Eventually, the situation became clearly unsustainable and the breakdown of collective farming was inevitable.

Getting land back to the peasants

With *Doi moi* ('renovation', economic reforms introduced in 1986), the household was again recognised as an economic unit and the farmers gradually gained control of their labour. For the rural families, the introduction of Decree 10 in 1988 signified the final change from collective farming to household production. This decree allowed the collectives to allocate the right for use of farmland to individual farming families (based on the number of family members) during a period of 10 to 20 years. Thereby, the households became free to organise and complete all farm work; they could use their share of the produce as they pleased, even sell it at free market prices, in order to improve the family economy (Kerkvliet 2005: 227–28). The household contract system – where a household was responsible for most stages in cultivation of land they had long-term use rights to and – initially increased agricultural production (ibid., Luong 2010: 192–93). As a result of this decollectivisation, quite a large amount of the labour was returned to the sphere of household work. Consequently, the household contract system (Decree 10) simultaneously contributed to a redirection of women back to the households and away from the public production of the cooperatives. Because the decree allowed the households to organise and manage their labour on the contracted fields, this opened up for a return of the male heads of households to have the final say about labour that the households had undertaken to perform – predominantly 'female' duties, like transplanting, weeding, and harvesting (Truong Thanh-Dam 1997: 86–87, White 1988: 175). Traditionally male duties, such as ploughing, harrowing, preparing seedlings and irrigation of the fields were still done collectively (Kerkvliet 2005: 184, 188).

In 1993, the National Assembly introduced a revised land law. The former, from 1987, had given the farming cooperatives the right to assign pieces of land to households for their own economic purposes, and stipulated that the allocation should be 'more or less equal'; even if the households did not own the land, they could use it for a 'prolonged period' (ibid.: 227). The new (1993) law stipulated the right to use fields with perennial crops for a period of 20 years. Around this time, many areas carried out a redistribution of arable land to the households (ibid.: 227–28). The right to use residential land did not have a time limit and forestland could be used for 50 years. According to the 1993 law, all members of peasant households, regardless of sex, were entitled to land use rights. Men and women were to have equal opportunities to receive and make use of their land rights, even though all land exclusively was administrated by the state. Eventually, land use certificates were issued to all the households, listing all the landholdings, and stating the name of the household head, which was usually a man. Another novelty this land law produced was the so called 'five rights', i.e. the right of 'family households and individuals receiving land from the State' to exchange, transfer (including inheritance and the right to sell the user right), rent, mortgage or give away their use rights (Ha Thi Phuong Tien 1997: 57).[10]

The present land situation is historically unprecedented in Vietnam as it combines time-limited allocation of individual/household rights to land use with ultimate state ownership. This move led to a rather substantial decline in poverty because of the increased agricultural output (National Human Development report 2001: 29). This improvement has levelled out after the initial great rise. In lack of a private land market, a market in land use rights has developed, enabled by the right to trans-

10 This means that private land ownership is prohibited and the state ultimately manages all land. 'Land belongs to the entire people with the State acting as the owner's representative and uniformly managing land', Land Law (No. 45/2013/QH13, Chapter 1, Article 4). The basic stipulations of the 1993 Land Law, i.e. individual/household rights to time-limited use rights to cultivated land and the 'five rights', are still valid, even if the law has been revised several times after that.

fer land rights. This has in turn led to a situation where peasants again can be landless, a crude fact for those peasants who – for reasons such as poverty or state expropriation – have been forced to sell their land rights (Tran Thi Van Anh 1999: 105).

Good peasants are few peasants

In the mid 1990s the state fervently wished to increase industrialisation and modernisation in the countryside. In order to achieve this, the engagement of rural households in non-farming activities was seen as the most desirable development. Contrary to the earlier policies, it was hoped that the Land Law was going to result in a process of concentrating land into the hands of a few large-scale and specialised farms. However, this has not yet happened on a large scale (Nguyen Sinh Cuc 1995: 113, 117). As voiced by Le Kha Phieu, the General Secretary of the Central Committee of the Communist Party in 2001, the desire was that the peasants who do not develop into sizeable agricultural producers should become 'workers' and 'intellectuals' (Le Kha Phieu 2001: 107). Following this ideal development, rural areas were supposed to gradually be urbanised and 'the Vietnamese peasantry will step by step get rid of poverty and backwardness.' (ibid.). But lack of capital, higher education and non-farm work, most of the farming households hung on to what little land use rights they had. Therefore, most of the farming is still household based and generates only a limited amount of surplus produce. This does not mean that farmers have missed the opportunities brought forth by a more open market. Most of the farming families in the village of my fieldwork had experienced quite extensive improvements of their economic and material standard over the last 15 to 20 years. Apart from improved farming outputs, many gave evidence of the major upgrading of the houses in the village. Previously, housing mainly consisted of cottages made of wood, rice straw, mud and lime, with thatched roofs; now concrete houses that can stand against the weather are more common, and most households also have some electrical equipment and devices. Nevertheless, the gap between the village and such cities as Hanoi was noticeable. The relatively high positions and prestige that the peasants received during revolution and warfare have not lasted into the present-day society. Increased demands on for example monetary success and technological advancement have been unattainable for many peasant households.

In post *Doi moi* Vietnam, the government still has a somewhat paternalistic attitude toward rural people, visible not least in official development reports, where they are pictured 'as poor, backward, remote, unconnected, unaware, and dependant on the state for their uplift' (Taylor 2007: 6). The heroic connotations of manual farm labour are largely gone, but the state's urge to encircle peasants with ideologically infused movements and policies is still there (e.g. McElwee 2007: 90–93; Taylor 2007: 7–12). Many villagers in Lang Xanh saw themselves as poor, primarily economically poor, but in some sense also culturally. This cultural poverty ranged from a lack of scientific training to 'refined' behaviour and taste. The inhabitants perceived the confinement to manual household farming as the major reason for this. The farmers' concern of appearing as outmoded was enforced by the regime's desire to be a scientific and technologically based country that had advanced from the stage of small-scale agrarian society (Taylor 2007: 9–10). Furthermore, the same ambition is mirrored in the state's elevation of agricultural methods that increase production by technologically controlling nature and the environment. Following Marxist ideas, the government and its development planners promote science and technology also as a means of dissociation from the spiritual relations that used to exist between farmers and farmland. Nature's conditions for farming are to be controlled by advancing techniques and any seeking of spiritual aid is disclaimed. Although reformation and collectivisation made a great rupture in the relationship between farmers and land, and new varieties of rice and pesticides were introduced, to some degree, the farmers still have a relation with the spirits and deities that they believe inhabit the environment (ibid.: 10–12), as shall be explored further in chapter 3.

Claims on land and the labour of 'women' and 'peasants'

The absolute rule of the government meant that it had complete power over the definition and use of land and places, which in reality meant the ability to control where and with what the farmers should work (see also Malarney 2002: 23–24). Other important aspects of place in rural society, such as patrilineal inheritance of residential land and homes and a strong tradition of patrilocality, attracted much less interest from the communist regime. By exercising total land control, the state primarily wanted to lay claim on the farmers' labour. As we have seen, the state defined and redefined the content and meaning of farm work depending on its needs and goals. From the farmers' perspective, labour became their most important – almost their only – asset, because this was their means to sustain themselves and their families in their restricted land situation. Their contribution of farm labour was the ideological and practical prerequisite for the peasant population's place as state members, and when farmland was allocated to the farmers after the collectives were dismantled, it was their engagement in farm labour that entitled the households to receive land use rights for fields.

Persisting gendered division of labour

Due to the commanding nature of the societal transformations implemented by the communist regime, the power that ideology had promised to peasants, and even more so to women, was in fact held by the state. Peasants and women had been depicted as categories that – with land redistribution and collective labour as means – should be elevated from earlier low positions in society. However, from an ideological as well as a practical perspective, gender equality and fighting against traditional labour division have all along been set aside, from the time of collective farming (1955–57 to 1981–88), to the period of economic renovation (starting 1986) and on. Despite an explicit ideology of gender equality, many of the traditional ideas about the value and proper division of male and female labour lingered on after collectivisation. This could go on partly because the tradition of a rather autonomous village authority had been adopted into the collectives. For instance, it was not uncommon that the men, and the much fewer women, within the local political and administrative structures were related to each other, or gave advantage to their kin and friends. Quite often, the leaders were practicing a distinct division between state policies and plans on the one hand and the actually running of the activities in the village on the other. The continued presence of a traditional family economy also contributed to preserve a patriarchal power structure. This was based largely on the gardens and tiny plots of land that the households could cultivate outside the agricultural collectives. The cultivation of these plots was organised by the male household heads – even though women did most of the work – in parallel to the work in the collectives (Wiegersma 1988: 162–71).

This land, which was supposed to be only supplementary and eventually unnecessary for the collective economy, was actually sometimes providing for more than half of the income of the households. Hence, the household economy continued to be largely of the traditional kind, and the traditional patriarchal structures of the families and villages were largely preserved (ibid.: 154–5, 166). An interesting point to note is that during the period of collective farming, many of the activities that were supposed to continue within the households and in the realm of the family economy, were commonly tasks associated with women's duties; for example, small-scale husbandry and vegetable growing. Moreover, women's labour in the agricultural collectives was often undervalued by male leaders and, in terms of equal labour division, it was more common that women took on traditionally 'male' farm duties, than *vice versa*. This was enhanced by the fact that the peek of collective farming chiefly coincided with the war against USA. This left much of the farming, and quite a lot of the managing, to the women, while the men fought at the warfront. However, when the men returned home, the gendered division of labour tended to fall back into conventional patterns, and women were again perceived as best suited for 'smaller', 'female' duties (see also Bradley 1989: 47, 48).

It can be concluded that much changed with the revolution and during the times of collective farming, but the status of women had not fundamentally changed. The interest of the state in gaining control of and directing the peasants' farm labour to fulfil national and political goals overshadowed the ideological idea of a gender-equal labour division. Furthermore, after *Doi moi* there was a trade-off between gender equality and economic efficiency because when national liberation and reunification were attained, striving for economic achievements took the upper hand. From the perspective of farming and labour division, this development meant a reduction of ideologically informed policies involving social and gender equality. Instead, there was an intensification of market-driven individual and household sustenance where each household organised and divided labour freely.

State definitions of the roles of women and peasants

Even though farm labour has, as mentioned above, lost much of its lustre, extensive manual labour in the fields is still an integral part of the lives of present-day Lang Xanh women. As we have seen, female farm labour has contributed immensely to feeding the country, and rural women of contemporary Vietnam have frequently been reminded that they are the allegedly proud inheritors of the national female tradition of intense work. From this perspective, Lang Xanh women were part of a traditionally recognised femininity of hard-working and morally respectable women. In creating images of national identity, both on the international and the domestic scene, definitions and depictions of 'proper' Vietnamese femininity have been central (Nguyen-vo Thu-huong 2008: 44; Pettus 2003: 4–15).

The state's defining and redefining femininity has lead to contradictory and opposing feminine ideals. While the communist leadership has proclaimed eradication of gender differences and elevation of the status of Vietnamese women, they have simultaneously expressed ideas – fully in line with Confucian doctrine – concerning women's inherent capacities as mothers and their 'natural' aptitude to selflessness, faithfulness and industriousness. Female emancipation has been a duty as much as a privilege when Vietnamese women have been instructed to enlighten themselves and in that way contest such diverse properties as 'backwardness', patriarchy, ignorance, and immorality. The right of the state to define female labour and characteristics, depending on time and purpose, has led to complex, contradictive and elusive feminine ideals (Pettus 2003; see also Gal and Kligman 2000: 4–6). Vietnamese men, or masculinity, has not been defined and singled out as a special category in need of transformation. The Confucian-supported patriarchy was officially condemned by the communist leaders, but it was never really replaced by a different male ideal. Men could be landlords, peasants, workers, soldiers etc. and in these capacities they could be either embraced or perceived as in need of change. However, they were – and still are – not only 'men'.

In many respects, 'women' and 'peasants' have shared similar fates. They have been portrayed as ardent fighters, loyally sacrificing their individual well-being for a larger cause. At other times, they have been the outmoded masses, burdensome for the nation unless they adjust properly. They have been pictured as backwards, conservative, and oppressed, but also as ideal citizens and the backbone of society. To be a 'female peasant' is thus to be one of the categories most thoroughly targeted by state ideology. Consequently, 'female' and 'peasant' identities in the village of Lang Xanh were created in relation to historically, explicit, and plentiful state ideologies and everyone was influenced by, and had to relate to, the compelling discourses surrounding these two categories. Clearly, men were not exempted from the pressures of ideology. I will return to this in the following chapters, but as far as state movements, campaigns, and overt ideological aims and definitions go, the categories of women and peasants have been overrepresented.

From an ideological and historically informed framework, the focus will now be turned to the local context, i.e. Lang Xanh and some facets of village life.

CHAPTER 2

The village of Lang Xanh

Vinh, a close friend of my host family, told us that if we had visited Lang Xanh as recently as the 1990s, most of the houses that were now made of concrete with cemented yards would have been cottages with thatched walls and roofs, and with dirt yards in front. Now, when making a right-turn from the main road leading to the district town, and after travelling the nearly 130 kilometres from Hanoi up to the northwest, into the district of Thanh Ba in Phu Tho province, a small cluster of mainly white or yellow houses can be detected among the lush green foliage. Like so many villages, Lang Xanh is surrounded by rice fields and, as a first sign of a more upland terrain, some rounded hills encircle the village. In contrast to the archetypical picture of the northern Vietnamese village, Lang Xanh do not consist of narrow alleys and houses crammed together, and is not bounded by a bamboo hedge. Instead, the houses are spread out rather sparsely, even if many of the residential plots are adjacent to each other. The houses are connected by a web of dirt roads and tracks, of which the main ones were remade in concrete during my stay. With few exceptions, the plots of the residential houses are enclosed by fences, walls, or hedges. Inside, you often find chickens picking about, maybe a pigsty, some vegetable beds, fruit trees and decorative plants. Most likely, all this is guarded by at least one ferocious dog – or so it seems to me.

The 'we' who arrived at Lang Xanh was I, together with the young woman, Phuong, who was my field assistant cum interpreter throughout the whole stay in the village. We had met in Hanoi where she had studied English and it turned out that she was from the province capital of Viet Tri. She was an invaluable bridge between the villagers and myself. She helped me, not only with language matters, but also when discussing and pondering over the things that we experienced together, although with different preconditions. We were virtually joined at the hip and when one of us appeared without the other, people immediately asked where the missing one was. You could say that many of the 'I's in the book to some extent imply a 'we'.

The choice of field site for this study was a collaborative effort. Earlier on, I had been part of a project that had included doing fieldwork in four villages in Phu Tho province, so some contacts had been established already. Moreover, the province was not too close to Hanoi, where the villages tended to be larger and with more diverse occupations, and it was not too far away from the Red River delta where the conditions for wet rice cultivations were poorer, which in turn led to a greater mix of sources of income, but for different reasons. I also wanted to find a village that consisted mainly of *Kinh*[1] inhabitants. It was in one of the central districts of this province that I conducted this fieldwork, from October 2002 until October 2003. With help from the Institute of Anthropology at the National Centre for Social Sciences and Humanities in Hanoi that introduced me to persons at the provincial level, it finally boiled down to the village, which I call Lang Xanh.

Lang Xanh is the centre of the commune, which consists of 12 villages. From a strict administrative viewpoint, a village is not an administrative unit in Vietnam (Luong 2010: 8, Malarney 2002: 13). The lowest administrative level is the commune (*xa*) and in most cases, the communes are made up of villages (*lang*) that the inhabitants perceive and refer to as separate communities, and this was also the case here. The commune's administrative centre was situated in Lang Xanh and most likely that was why this village was home to more cadres and workers than the other villages, though the majority of the inhabitants engaged in farming. These circumstances, combined with the relatively flat fields that belonged to Lang Xanh, made people in this village appear – to themselves as well as their neighbours – to spend comparatively less time toiling the fields and to have more time to engage in 'emotional' (*tinh cam*) things, like socialising, singing, meetings or just watching TV.

1 *Kinh* is the term for the majority population in Vietnam, i.e. the ethnically Vietnamese, or Viet, people. They constitute around 87 per cent of the population. Apart from the *Kinh*, there are 54 recognised ethnic minority groups represented in Vietnam – see http: //www.chinhphu.vn/portal/page/portal/English/TheSocialistRepublicOfVietnam/AboutVietnam/AboutVietnamDetail?categoryId=10000103&articleId=10002652.

The other villages in the commune were described as being situated 'in the forest', which nowadays is more a figurative than a literal description.

Fields and farming

Lang Xanh, which grows rice as the main crop, is some distance away from the most fertile soils of the Red River delta. This means that the land quality is reasonable even if not the best and is probably one of the reasons why the houses were less densely placed compared to the villages in the richest delta zones, where as much land as possible is used for growing rice. As far as land quality is concerned, the Mekong delta in the south of Vietnam has the most excellent conditions for wet-rice cultivation, and the villages are much denser.

Enfolded by the relatively few and low hills, and divided up by the roads that led to neighbouring villages, were six major areas of rice fields. Within each of these areas, each field was separated from the others by a network of banks that enabled the farmers to regulate the water level according to seasonal requirements. The size of the fields varied quite a lot. The larger ones were slightly more than one *sao*[2], while the smallest fields were just a couple of square meters. A household could have fields in all six areas, though many had their fields more gathered together. The distribution of the fields shall be discussed in more detail below. The dispersion of the fields resulted, however, in a lot of transportation of tools, fertilizers, rice seedlings and harvested crops. When the seasons so demanded, the roads and paths in and around the village were filled with men, and even more so, with women, who carried things to and from the fields with the help of shoulder poles, bicycles, and carts pulled by persons, cows, or buffalos.

Some of the fields had sour soil and were so muddy that they virtually had no bottom, and some were obscured by the hills, preventing them from getting as much sunshine as they would ideally need. Nevertheless, the rice fields in general gave two crops of rice per year, and between the summer and the winter crop, i.e. October to January, virtually all households grew a crop of corn (Zea mays). The hills were, however, not merely an obstacle for the desired sunshine, but were also used for growing manioc[3] – which was, just like corn, a fairly recent crop, and also

2 1 *sao* is 360 square meters.

3 Manoic is a starchy tuber, 15–40 cm long, with white flesh and rough, brown, rind.

primarily intended for animals. Some tea bushes, and a few patches of still quite young trees could also be seen. These hills were, to my initial puzzlement, often referred to as 'the forest'. Since there were not that many trees outside of Lang Xanh, except for the scattered odd plots of tiny eucalyptus-like trees, I was somewhat confused when people I met on the way to the hills commented that apparently I was heading for the forest. My landlord and one of his frequently visiting friends explained that the village used to be surrounded by lush areas of bamboo, like a forest. They both remembered how they as young boys could go out and pick nice and juicy bamboo shoots anywhere around the village. Now they had to be bought at the market. The bamboo was cut down, they said, to give way to manioc which people needed more. When food was scarce, in wartime and during collective farming, it had been necessary to grow as much as possible. The trees had been more plentiful then too, but farmers had cut them down to sell to the forest enterprises needing wood for the paper factories. Now, as then, people tried to make the most of the land at their disposal and there were few if any patches that were not used for growing crops.

When manioc was planted, in late February and early March, it was not uncommon to see whole families up on the hills. The situation was very similar with corn. Both men and women maintained that manioc and corn cultivation was incredibly easy. No particular skills were necessary and anyone, regardless of sex and age, could do it. This was evident by the fact that women and men quite often performed the same chores and quite often performed them together, even if women generally worked more than men with these tasks. Seemingly none of the chores of manioc and corn growing were perceived as especially delicate or feminine. It was coarse and hard labour and, thus, a duty where most of the household members could contribute in one way or the other. With rice, the duties were much more strictly separated, as we shall see in the following chapters.

The importance of rice

Of all the farmland in the village, the rice fields were the most highly valued type of land. When asked about their land holdings, all villager answered by stating the area of rice fields at their disposal. The same happened in discussions about fields that people had passed on to their grown-up and married children, or other farmland transactions that the households were involved in – it was all about rice fields. A family often stated its level of sustenance by saying for how many months per year they lacked rice. The poorer households in Lang Xanh normally lacked rice for a couple of months every year, which meant that the rice they grew themselves was not enough to feed them all year around, but had to be supplemented by bought rice. These families often had to sell their labour to cover this shortage. Even though the villagers were part of a monetary and commercial economy, the idea that a sufficient supply of rice was essentially the same thing as affluence still lived on. 'Rice is money' as my landlord expressed it when he told me that tax, which is now payable in money, used to be paid in rice. The price of rice was, however, taken into consideration when taxes to the commune were set. Having sufficient amounts of rice conveyed a feeling of security, you knew that you would always have something to eat, and the consumption of home grown rice also ensured that the rice was of good quality and that it was clean and not full of pesticides. The villagers assured me that the sellers always kept the best rice for themselves.

In many ways, rice cultivation was the quintessence of farming in Lang Xanh and as such it was well engraved in its history and tradition. Vietnam is firmly anchored in Asia's rice belt, which stretches from India all the way to Japan. Within this area the cultures and environments are many and diverse, but through the centrality and common devotion to rice, it is still possible to detect a kind of 'rice culture' that connects the region (Hamilton 2003: 23). Cultural practices encompassing rituals, organisations of communal labour, arts, and language all point to the importance of rice cultivation. In a global perspective, rice is the crop that occupies the largest area of farmland and it provides more consumed calories than any other food (ibid.). In places where rice cultivation dominates, the different stages in the agricultural year often greatly influence the cycle of activity, with rituals marking the most important turning points, like transplant and harvest (ibid.: 30).

This was, and to some extent still is, clearly reflected in Vietnamese annual rituals. *Tet*, the Vietnamese New Year, is, for example, celebrated in the less intensive period when the last rice crop of the year is harvested and before the transplanting of the first rice crop of the new year. Asia's 'green revolution', which began in the late 1960s, introduced new, more hardy and high-yielding varieties that ripened quicker, thereby enabling more crops per year in several areas. This affected the rhythm of the agricultural year, which sometimes got out of step with the ritual cycle (ibid.: 22, 439). In the case of Vietnam, the rituals involved with rice cultivation were also diminished by the communist regime's aversion to rituals involving supernatural components, something that chapter 3 will further touch upon. Nonetheless, some households in Lang Xanh still made offerings of newly harvested rice (*com moi*) to the deities and the ancestors. This normally happened on the 5th of lunar May,[4] on *Tet Doan Ngo,* the so-called parasite-killing festival, when offerings were made to the spirit world to avoid pestilence and disease, and on the 10th of lunar October. One male farmer, a man in his 50s, told me that his family offered new rice when they felt happy, healthy and at ease. However, if they were very busy or less content, they refrained from making an offering that year. According to many of the villagers, the offering of new rice is an old tradition that you do not have to follow anymore. And, as just mentioned, the traditional agricultural calendar is not followed very strictly. With the introduction of new rice types and a more individual organisation of labour, the dates for the offerings of the new rice do not really coincide with the rice harvest anymore. Moreover, the relatively newly introduced crop of corn, grown between the summer and the winter rice crops, meant that harvesting took place as early as possible, so most people had already eaten from the new harvest before the traditional dates of offering new rice.

Another common factor among the 'rice cultures' is that rice – besides being a necessity in a proper human meal – in all its forms and shapes is essential in the food offerings directed to ancestors, deities, and spirits (Hamilton 2003: 30). This belief was shared by the people

4 Like China, Vietnam has a lunar calendar (*am lich*) which is used parallel to the Western solar calendar (*duong lich*). Year one in the lunar calendar was in 2,637 B.C. and it was (and to some extent still is) used to regulate the farming year as well as festivals, rituals, ancestor worship and other spiritual activities. The lunar year has 12 months with 29 or 30 days, all together 355 days, based on the time it takes for the moon to complete a phasic cycle. In order to synchronize the lunar year with the solar year, an extra month is added between the third and the fourth lunar month approximately every third year according to the lunar calendar. In this book, if the name of a month is not indicated as, e.g. lunar May, it is the solar month that is intended.

in Lang Xanh as well. When the ancestors were offered their meal in accordance with the lunar calendar (more on this in chapter 3), they, like all the other otherworldly beings, were always presented with rice and rice liquor. As a complement to this, other food items could be added according to the supply and opinion of the household. In this respect, rice is not comparable to any other food item and in Vietnam the word for 'meal' is the same as the word for cooked rice '*com*'. Hence, the phrase 'have you eaten yet?' (*an com chua?*) literally translates into 'have you eaten cooked rice yet?'. This is a trait that many languages in rice-eating Asia have in common (ibid.).

Some farm chores, like transplanting, are distinctive of this type of agriculture (ibid.: 43, 51). In many respects, life in Lang Xanh was still characterised by rice cultivation, as the following chapters will demonstrate from various angles. In principle, the agricultural year started, as mentioned, with transplanting of the winter crop just after *Tet* (new year), i.e. in mid to late January, sometimes even in early February. Transplanting usually consisted of planting pre-grown rice seedlings in the water-filled fields. The seedlings were collected into bundles, brought to the field where the transplanters carried one bundle at a time, from which they planted little bouquets of seedlings into the water-filled mud. Occasionally, smaller fields could be sown through broadcasting, i.e. throwing non-germinated seed rice straight onto the fields, instead of transplanting seedlings. This method was quicker, but the plants needed to be thinned out and the seedlings were more vulnerable to insects and rodents when they were not pre-grown. At this time of the year, the weather was often damp and cold, sometimes not more than four degrees Celsius. Transplanting was the most 'female' duty of all farming chores, and the women worked barefoot in the cold and muddy water. In most fields, the water reached knee-level, but in some fields the mud was so soft that bamboo poles were laid out to walk on. Shoes or boots immediately got stuck, so bare feet were the only option. Sometimes the women wore double trousers and extra sleeves outside their shirts and jackets to protect arms and legs from the cold, but at the end of the day, their limbs were numb with cold. After two to three weeks, most of the fields were transplanted and it was time to plant manioc on the dry hill fields. At this time, most of the households also planted peanuts, sweet potatoes, beans and leafy greens. Just like rice, this was mainly for household consumption. From around mid February until the rice harvest in May, there was a slack in the farming season. During that period, many women engaged in raising pigs, some grew cat's ears (an appreciated mushroom) – an activity performed even more often by men – or tended to tea plants, which some households grew on the hills or in parts of their gardens. A few of the women traded in fruit and vegetables, either in Hanoi or at nearby markets. Many of the men worked as construction workers at sites in the vicinity of the village, or in other parts of the country, and only some of them came back for peak periods of farming. Most of the women who worked outside of the village returned at least to participate in transplanting and harvesting.

In early to mid May, it was time to harvest the rice fields. With very few exceptions, this was done by women who cut the rice and tied it into sheaves, and men who collected and carried them, sometimes by shoulder poles but mostly by carts, back to the houses where they were stored until they were threshed and husked. When the fields were harvested, a busy time followed for the very few women and many men who ploughed and harrowed the fields in preparation for the next rice crop. By the middle of June, the fields were again transplanted with bright green rice seedlings and now it was the persistent heat that was the greatest obstacle to the women gathered in the fields. They were covered in clothes and conical hats to protect themselves from the burning sun. When August came, after another less busy period in the farming season, it was time to harvest the rice and prepare the fields – this time for corn. From September until December (when the manioc was harvested), most of the rice fields were emptied of water and gave a crop of corn, before the cycle started all over again, with preparing the fields for the winter rice crop.

Chanh, a former teacher at the Agricultural College, told me that people here loved (*yeu*) the land. If they could not see to the rice fields or the manioc hills at least once a day, they would experience a growing frustration creeping up on them. The farmers in Lang Xanh were attached (*gan bo*) to the land; this relationship was inevitable in the same sense as with a spouse or a relative, which is what the phrase '*gan bo*' indicated.

Introducing the village

The People's Committee – the administrative and political structure

When entering Lang Xanh, one of the first sets of buildings to appear were the yellow rows of houses placed in a u-shape around a yard, which was the commune's People's Committee (*Uy Ban Nhan Dan* often abbreviated to UBND). This was where the Chairman, the vice-Chairman, and a handful of cadres of the commune had their offices and where meetings were held. They were in charge of everything from budgets, infrastructure, and resources, to the socio-economic development of the commune. If they were in, people could come and visit a cadre with their concerns and there were newspapers available in the main hall, free for anyone to read. However, most of the time it was rather quiet at the People's Committee.

The People's Committee's principal duty is to be the executive agency of the District People's Council and the Communal People's Council at a local level. Vietnam is a socialist state, lead by the Communist Party. The Party has a national structure that covers every level of political rule, from Hanoi, through the provinces and districts, down to the communes and villages. Parallel to this organisation – and more or less as a mirror image – runs the government's structure, where the People's Committee in the communes are the smallest units, which manages state affairs and implements the policies and decisions of the government and the National Assembly. Since the reformations starting in 1986, the still powerful Party organisation has decreased and do not directly participate in the work of the government (Dang Phong and Beresford 1998: 88). The Party and the government, with their networks of cadres are ubiquitously present, and so are the, formally, non-Party and non-government mass organisations (e.g. Women's Union, Veteran's Union and Youth Union) that are organised in a similar, nation-wide fashion. In practice, the liaisons between these three structures are tight. The Party, the government, and the mass organisations were all present in Lang Xanh and made their imprint on village life.

Accordingly, Lang Xanh was represented in the commune's Communist Party and there were branches of the Women's Union, Farmers' Union, Youth Union, Veterans' Union, and Old People's Union. Of these the Women's Union was the biggest with about 70 members. In the commune where Lang Xanh is situated, it was preferred that persons who held an official appointment should also be members of the Party. This, however, is not a national, formal requirement. All the unions and the Party had monthly meetings and virtually all of their activities were initiated and controlled according to a top-down principle. The leader of the commune's Farmers' Union told me that some of the goals were shared by most of the unions, like family planning and information about policies and regulations that are set by the Communist Party at the national level. For example, if a family transgressed the policy of having no more than two children, the unions that the parents were members of could be criticised and the leaders might be seen as deficient in their assignment as union leaders. And if a couple were planning to get divorced, the union leaders – especially of the Women's Union – or other cadres, were required to talk to the couple in order to make them try to stay together.

Also, the unions took an active part in celebrating the National Day on 2 September (when the declaration of independence was made in 1945). Then singing and dancing performances were held. The leader of the Women's Union in the commune told me that they had received instructions from higher levels in the organisation that this year they should have a performance together with the Veterans' Union to celebrate the National Day. Rehearsals had begun at the communal 'cultural house' (*nha van hoa*) at the People's Committee a few days before. The general idea was that the unions should not only convey governmental information and decisions, but also endorse cultural and recreational activities. The older and middle aged members anticipated the event – because, apart form the compulsory speeches held by officials from the People's Committee, it involved singing and dancing acts which they approached with some enthusiasm – and they had had quite a bit of fun during rehearsals. The much fewer younger (under the age of forty) members of the Women's Union did not participate in the performance.

The leader of Lang Xanh's division of the Women's Union had noticed a decline in the number of members. Young, newly married women with small children were not very interested in paying the yearly fee. They thought – not entirely unfounded – that the union was mostly about carrying out orders from district level. Many women were busy and did not want to spend the little free time they had on union activities. The woman who was the leader in the commune told me that she did not particularly want to become a Women's Union leader, but found it hard to refuse. I got the impression that the existence of the Women's Union, and the other mass organisations, was at least in part a duty rather than a sign of interest.

The man from the People's Committee that I met most often was Loi. He was a man approaching his forties and lived in Lang Xanh with his wife and two sons. As the vice-Chairman of the commune's Peoples Committee he was also in charge of agricultural matters. It was his responsibility to inform the village leaders in the commune about new ideas and methods in agriculture. He and his wife had rice fields to cultivate, though his commitments at the commune office kept him from actually performing much farm work, so therefore his wife managed most of it. According to him, many farmers were quite difficult to convince when it came to testing new ideas. He wanted to encourage the other farmers to try new types of seeds and to improve agricultural methods. The best method, Loi thought, was to show them how it worked in practice so that they could experience the outcome for themselves. He shared the common idea that it was much easier for women to try new things if they had their husbands' support and encouragement. If not, he thought, the wives were reluctant to change the way they worked the land. Of course, Loi added, that there were exceptions to be found and he mentioned my landlady as an example of an enterprising female farmer who was quick to try new things. Young people could also be rather conservative when it came to farming. According to Loi, they picked up most of their ideas from their parents in a rather unreflective manner, and the young people who stayed and became farmers in Lang Xanh were, in most cases, those who were not successful in their studies, and lacked the means to choose some other line of work.

So far, the commune had turned down inquiries from the district about turning agricultural land into industrial grounds. According to Loi, it was important to protect the rice fields so that the farmers had sustenance. He believed that 'the land situation should be stable so the farmers feel that it is worthwhile to invest time and money in the land.'

Just behind the People's Committee was a small building that contained two rooms. This building was also covered in the yellow plaster that can be found on many official buildings in Vietnam. This was the commune's police office, where Tu, the chief of police in the commune, together with a couple of colleagues, could be found.

Tu was initially very nervous about my presence in the village. He felt responsible for my safety, at the same time as he was anxious that I might pick up something unfavourable, or probe too deeply into possibly sensitive land issues. In the beginning of my fieldwork, he visited the household where I stayed several times, and he eagerly stressed the importance of a weekly plan of my whereabouts including whom I planned to talk to and about what. But when he had stated these requests he usually turned to his favourite topic in our conversations, which was gender (*gioi tinh*). He knew I was interested in studying gender relations and his thoughts on the matter revolved much around the possible differences and similarities between Vietnam and Sweden when it came to issues like courting and attractions between men and women. The jokes and speculations about these matters were frequent during my time in Lang Xanh, though they lessened over time. Compared to land issues, gender was a safe topic that did not invoke uneasiness or unrest among local authorities.

In the end, Tu settled with the request that I inform him when I left the village to go back to Hanoi, and also when I returned. This information was common knowledge to everyone in the village virtually at the same time as it happened, but Tu still wanted me to come and see him in connection with my comings and goings. These requests intensified when, towards the end of my stay, Vietnam was hosting the Southeast Asian Games and the general level of security increased. The most notable effect of this in Lang Xanh was that the handful of women who normally went to Hanoi to sell vegetables and fruit on the streets when farming was in low season, were not allowed to continue their activity since ambulating street vendors did not fit into what the state and the organisers of the games wanted to show to the visitors.

The village leader and some facts and figures about the village

Further into the village was the home of Binh. He had been the village leader just over three years when I arrived at Lang Xanh. His appointment did not render him an office at the People's Committee and, like the commune cadres, he lived in a private home. He was a thin man in his late forties and a member of the Communist Party and the Farmers' Union. Our first meeting took place in his house to which my landlords' mother had accompanied us in order to make sure that Binh's allegedly vicious dog did not eat us alive when we entered the premises. Binh told me that Party membership was a prerequisite for being a village leader. It was the people who had elected him, but the candidates where selected by the People's Committee in the commune, and the General Secretary of the local Communist Party usually had to give his approval before the villagers could elect their representative. The position of village leader was an intermediary role, between the villagers and the People's Committee, whom he was under the direction of. He too was somewhat nervous when I first installed myself in this village, not least because Tu, the chief of police, wanted him to report back every evening on whom I had seen and what questions I had asked people. Luckily, both for him and me, this idea quickly fell into oblivion. After Binh had accompanied me to one of the households and I had presented the main purpose of my study and asked some introductory questions related to farming and farm work, he came to the conclusion that following me around would be much too time consuming. Being a part time farmer himself and married to a full time farming woman, he already knew all he needed to know about gendered labour division, land access, and the organisation of farm work.

On our first encounter, the village leader shared some facts about the village with me. He told me that Lang Xanh consisted of 132 households comprising 460 persons. Of these, six were non-farming households and did not hold agricultural land. The adult members of these households supported themselves as teachers, factory workers, or fulltime cadres.

Many people came to live in Lang Xanh in the 1940s. This was then a sparsely populated area and the landscape was dominated by forest-covered hills. People who settled here came from more densely populated parts of the country, like the coastal province of Thai Binh, or from the township (*thi xa*) of Phu Tho, which was a likely target for the French soldiers during the war in the later part of the 1940s[5]. Since people had come to Lang Xanh from different areas there was a mix of patrilineages, none of which had ever been dominant. Generally, it is not unusual that one, or a few, powerful patrilineages with a large number of

5 Later on I learn from some more senior villagers that Lang Xanh was attacked twice by French soldiers and the residents had to hastily grab a few belongings and flee into the surrounding forest. Cows and buffalos were left to wander around freely. The French soldiers stole items and animals, and sometimes captured people. The war against the USA also affected the village. Some villagers left their houses to Vietnamese soldiers, and took to the then-thick forest again.

members exert quite extensive control over the doings at many levels in a village. Here, the village leader was keen to let me know, there were no conflicts and people helped each other, even if they were not related. No major income gaps between people from certain kin groups or places of origin could be noticed, something that is not uncommon in other villages, Binh informed me.

By this time, the original lineages had married into each other for some decades and many families had become related in this way. The myriads of connections between the households gradually revealed themselves to me, and were more intricate than I initially realised. In the beginning, when I started to get acquainted with people, I was surprised that people had not told me that their neighbour or co-worker was also, for instance, a sister or an aunt or a brother-in-law. For most of the villagers these links were so taken for granted that they were usually not mentioned. The qualities of being a relative, a neighbour, and a friend were quite often merged into one and the same person. After a couple of weeks I started to make a map over the village and draw lines with different colours indicating kin relations between the households. My host family found this amusing and interesting and we spent quite some time elaborating on the connections and museing over the mishmash of lines that spiralled on the map.

Apart from some in-and-out-moving spouses, the population has been rather stable for the last decades, according to the village leader. After the war against USA ended in 1975, some households moved to the south of the country. Most of these households were founded by soldiers who were stationed here during the war, and when it was possible, they moved back again. But some of them stayed, usually because they had married a local woman.

From the village leader I learnt that there was only one female-headed household in Lang Xanh. This was rather unusual, I believed, but Binh said that this woman was divorced and her parents were dead, so she had her own household. There were four unmarried women who had children, but according to the village leader, they all lived with their parents. Quite soon, however, we met several women who, for one reason or another, lived without a husband and most of them did not share their house with a parent. It also turned out that the woman Binh was referring to as divorced never had been married but brought up her child on her own. As it turned out, this was also the case with at least eight other women in Lang Xanh. These women originated from the village, or the surrounding neighbourhoods, and had parents or other relatives here. From a more official standpoint, these households were preferably seen as offshoots of their parents' households, rather than separate female-headed households.

Binh told me that most of the elderly villagers lived with their children. If the children were working far away for some period, the grandchildren stayed with the elderly. During my year in the village, I only met two old men and one woman who did not fit into this description. Despite their age, they preferred to live alone and do things in their own way, even if their children lived in the village and could give them a hand when it was needed. 'Old people can be difficult', as one of the old men later on said to me, maybe in a moment of self-knowledge or perhaps careful diplomacy. The village leader emphasised that it was not uncommon for three and sometimes even four generations to share a house. His remark was held out as an example of a caring family and village atmosphere – a highly valued traditional ideal in the official discourse, in line with the criteria for a 'cultural village'. To achieve the status of cultural village there were numerous criteria – including material, economic, educational and moral development – that should be met. The focus was on preserving and promoting 'fine' cultural traditions (e.g. harmonious and respectful relations within households, families, and the village), while opposing 'backward' customs (like superstition, and elaborate weddings and funerals). Promoting children's education and health, applying family planning (preferably no more than two children) and avoiding divorces were other criteria. Another important point was to combat 'social evils', like drugs, gambling and prostitution (see also Luong 2010: 242–43; Werner 2009: 66).

When it was time for a village meeting, the village leader was the one who summoned the meeting. To gather the people, he beat the large drum that he kept at his house. The meetings occurred irregularly. There could be two meetings in one month, or there could be several months between the meetings. Causes for village meetings could be that some new information had come from higher levels in the administrative structure, like new decrees or fees to be paid, or information concerning insect control, road constructions, or small husbandry projects for example. The number of participants in the village meetings tended to be quite small and the partakers were of relatively high average age. Other duties the village leader carried out was to collect tax from the farmers and hand it over to the cooperative or the People's Committee and – again by beating the drum – gather men together to dig a grave when a villager has passed away.

As far as the health situation was concerned there was a health care station in the commune, where one doctor, four nurses, and a part time specialist in reproductive health, worked. There was also a nurse in each village in the commune. The nurses' main duty was to spread information to the households about sanitation, including mosquito protection. Other facilities that were available to the people in Lang Xanh were a kindergarten, a primary school, and a secondary school. Almost all the children went through nine years of schooling, starting at the age of seven. College and high school attendance were lower, and the students had to travel out of the commune to get to these schools. So far, only a handful of the young people had gone to university.

The last facts about Lang Xanh that the village leader provided was that nine households had a telephone line while all had electricity and access to wells for their water supply. Slightly more than half of the households owned a motorbike and the rest at least had a bicycle. About seven months into my fieldwork, *thi xa* Phu Tho erected a telephone mast in connection with a jubilee, enabling the use of mobile phones.

When we were about to leave his house after our first meeting, the village leader remarked that even though the population was relatively stable, the village could be rather quiet when there were no festivals or special occasions, such as weddings or funerals. When things were going on a usual, it was the youngest, the oldest, and women who dominated village life. The youth were busy studying, and some had left, at least temporarily, to learn a trade or earn money. Quite a lot of the men went away to work, at constructions sites or similar, often far from home. When farming was in low season, a few of the women also left to engage in market activities, mainly selling fruit and vegetables. According to Binh, this set a mark on the atmosphere of the village, something we shall come back to in a subsequent chapter.

My host family – multi-generational residency

The house where my host family lived was not far from the village leader's home, and belonged to the 'high-quality' range housing having concrete walls and a paved courtyard. The garden, which included papaya, mango, and banana trees as well as some beautiful flowers, sloped towards a stream. What made this house stick out, though, was the fact that it had a small tiled room with a WC. This extraordinary facility had been installed when the eldest son met a woman from Hanoi, whom he married during my stay with the family. When he came to visit together with his girlfriend, she would thus not have to use a concrete pit adjacent to the pigsty; an arrangement that was common in most of the other houses.

My landlord, Minh, had grown up in this house, which used to be his parents' home, and his elderly, widowed mother had now come back to live with her eldest son and his family after having lived with her middle son for a couple of years. The house had once been the home of her parents-in-law and the grave of her mother-in-law was situated in the garden, not far from the fishpond. The custom of burying family members in the garden, or in the rice fields, had come to an end when the People's Committee had designated a hillside in the outskirts of the village as the official graveyard. This was where her husband and her father-in-law had their graves and where everyone who passed away in the village nowadays was buried. One of the main purposes for creating communal graveyards was to lessen the distinctions in wealth and status that the graves used to indicate, either by their impressive headstones or even small mausoleums – or by the lack of them (see also Malarney 2002: 137). Sometimes geomancers used to be consulted by relatively well-off families to find the most prosperous direction for a grave, to be placed in gardens, rice fields or in relatively public places in the village. Occasionally, it still happened that someone in the village contacted a geomancer in order to find the best and most propitious plot at the graveyard. The building of mausoleums had ceased, however, and the differences between the headstones were not too striking.

My landlady, Lan, was somewhat bothered by the grave in the garden, she found it old-fashioned and possibly unhygienic, but her mother-in-law did not want to hear about moving it. This house with its garden was incorporated into the family line and Minh hoped that it would continue to be so, either through one of his sons or his brother's sons.

The cadastral officer and land concerns

In the flattest and probably most densely populated area of Lang Xanh lived Han, his wife, and their two teenage sons. My first meeting with

Han, the cadastral officer in the commune, almost never came to be. Since he was in charge of the land administration in the commune, I was eager to see him as soon as possible and we had agreed to meet at the People's Committee at 8 o'clock on a cold, damp, and grey Wednesday morning. After one hour's waiting and no sign of Han, we decided to go to his house, which was not very far away. We arrived in the middle of his breakfast, and since this was only a few days after *Tet* (the much celebrated Vietnamese new year) we were treated to sweets, fruits and some liquor. This was probably a more pleasant and relaxed atmosphere for a first meeting, and we also met Han's wife, a farming woman whom we eventually became much more acquainted with than with her husband. On our first encounter, I learned that on unpleasant mornings like this one, there was no need to rush to work.

Land allocation

The processes of land allocation in this village had, according to Han, taken place approximately in the following way: In 1980, when collective farming was declining, some farmland was distributed to each household in accordance with the number of farming labourers and dependent persons in the households. This was a decree from the Communist Party, he explained[6]. In 1988, there was another allocation of farmland[7], this time more extensive than the former, he added, but people still worked under the authority of the cooperative and the state. Of late, farmers are allowed to work more flexibly and to have more freedom he said, referring to the 1993 Land Law giving farming households the right to cultivate their fields according to their own decisions but, with full responsibility for the outcome. Each household was now free to sell, mortgage, lend, or rent out the use rights to their fields. On average, each person had about one *sao* of allocated wet-rice field. The amount could vary according to the quality of the land. For the last two decades, there had been very few changes in the land holdings of the households, and it turned out that

6 This corresponded to Decree 100 that was issued by the Party in early 1981. This decree stated that leaders of farming cooperatives could contract workers to perform and be responsible for some defined stages in farming, as part of the collective farm (Kerkvliet 2005: 183–84).

7 This allocation corresponds to Decree 10 (1988). This decree made it officially possible to phase-out collective farming (see e.g. Kerkvliet 2005: 227–28). See also chapter 1 for more details.

the latest allocation, the one initiated by the 1993 Land Law, had been more of an update rather than a reallocation of the households' access to farmland. Some corrections had been made in relation to the number of persons, regardless of age and sex, who currently lived in the household.

At the time of the most extensive land allocation (in 1988), a few of the households had been indebted to the farming cooperative, because they had taken out more resources than their earned points had covered. The basic idea of the farming cooperative was that each worker accumulated work points corresponding to the type of labour she or he did and how many working days they had contributed. The workers were then remunerated accordingly. Remunerations were predominantly made in rice, but other collectively produced crops, fish or meat, for example, could be used as payment. In addition, the number of children and elderly in a household were accounted for when remunerations were calculated.[8] Huu, a woman of 37, later told me that when the farming cooperative was dissolved and land was distributed to their household, her husband was indebted to the cooperative because he had taken out produce in advance to pay for his gambling and drinking habits. Such debts had been balanced in the land allocation, with the result that their households received a relatively smaller land area. She was now divorced and lived with her teenage daughter and her old adoptive mother. Her household still had comparably few fields and she did not have the means to rent any additional land. Nonetheless, taking this type of differences into account, the diversities in land access between the households (counted on land per head in each household) were not very significant. Besides, there were few households either moving into or out of the village and most of the (mainly informal) land transactions thus took place between the long-term residents of the village in accordance with changes in circumstances and life-cycle situation. When a household member moved out, there were no compulsory changes made to that particular household's quantity of farmland.

Newly set-up households, the cadastral officer explained, received land from their parents, this was because all the farmland was allocated

8 The distribution of goods to members of cooperatives was quite a complicated affair. See Kerkvliet 2005: 55–59 for an account of how it could work...

and there would not be a new allocation until maybe in 2013,[9] i.e. 20 years after the 1993 Land Law was created. People had to share their land to make the best of the situation, but the land shortage was not great, he assured me. In retrospect, I believe that in a sense most villagers would agree, but all of them wanted more land so that they had better means to improve their situation. Grown-up children, both sons and daughters, were given fields from their parents when they married and started families of their own. But, Han added, all changes in land holdings had to be made in accordance with the Land Law. When villagers sold fields to get some money both the husband and the wife of the household had to agree to the transaction, as the law stipulated. But the only name that was set out in the Red Book (the land use certificate that each household had) was the name of the head of household, and in most cases that was the husband. All the households of the village had been provided with a Red Book by December 1998. It had taken five years to get them issued.

The Red Book was the tangible and theoretical proof of a household's land holdings, but in practice, it was not the land use certificates that informed people of which field belonged to which household. Instead, this information was protected by shared common knowledge among the villagers, farming as well as non-farming. Even the teachers, the cadres, and the children all knew which field was used by whom. My landlady Lan, for instance, had changed, bought, and sold many fields throughout her years as a farmer, but did not feel the least worried that somebody else would lay claim to any of her current fields. She had the temporary use right to several fields that were not listed in her Red Book, but she never for one moment thought that someone else might take possession of a field because she could not prove with her certificate that it was hers. She confidently told me that everybody knew which fields she used and she had never paid any attention to her Red Book. She did not even know which land was listed in it, and it appeared that in her particular case, the residential land in her Red Book was the piece of land which held their old house that had been sold some years

9 A revised Land Law which prolonged the use rights of agricultural land from 20 to 50 years came into effect on 1 July 2014. See Law No. 45/2013/QH13 Land Law, Chapter 10, Section 1, Article 126.2. This means that the reallocation of farmland that the villagers of Lang Xanh anticipated has not happened.

ago, when they were moving in with her mother-in-law in her husband's childhood home. The certificate that listed their present residential land was in the name of her mother-in-law, and her second youngest brother-in-law kept the certificate because he and another brother, with his wife and children, were living there at the time when the land was allocated. The fields allocated to her three brothers-in-law had, by now, changed users within the family many times, and none of the Red Books in the family of her in-laws listed the fields in accordance with their current users. Eventually the fields would probably be reallocated anyway, she remarked light-heartedly.

Renting fields

One way for households to get access to some extra agricultural land was to rent land from a family who were willing, or needed to, rent out some of their fields. Rice was the common currency for rice field-rentals and the agreements were made from crop to crop. Sometimes contracts were drawn up, but in most cases such affairs were settled by oral agreement. If, for example, a household's children left the village in pursuit of a non-farming career, renting out spare fields to people in greater need of them was a way to manage the situation. In this way, the parents did not have to give up the use right permanently, and they would get some extra rice as rent. People who rented fields were often farmers who had formed their households, or had children, after the allocation of land had been completed.

If a household wanted to sell their use right to a field permanently, it was not difficult to find buyers. This sometimes happened, but buying fields was only an option for the more well-to-do households. Good fields were expensive and the sellers were few, and this kept the prices up. In sum, it can be said that farming families who had many members at the time of the land distribution but had experienced a decrease in household members in the following years, had benefited from the allocation in a way that other families had not.

Land access

When it came to the actual procedures of land allocation in the village, the cadastral officer was, as I experienced it, rather restrictive with information. In due course, however, the picture was gradually made clearer for me as the villagers filled in more and more of the blanks.

Many of the farmers told me – and this was officially confirmed through Loi (the vice-Chairman of the commune's People's Committee) – that the first land allocation, in the early 1980s, had been accomplished by staging a kind of lottery. This meant that one person from each household had to draw numbered tickets equivalent to the amount of farmland that that particular household was entitled to. The number on each ticket corresponded to a cadastral map that showed the numbered fields. Apparently it was the villagers themselves who had requested this procedure to make it as fair as possible. This of course resulted in landholdings that were spread out in several directions around the village. What had puzzled me for some time, was the fact that quite a few households I had talked to seemed to have most of their fields in the same area and sometimes even adjacent to each other. It did not seem plausible that this could have happened by drawing 'land tickets'. The explanation was, of course, that immediately after the lottery was completed, a large number of families had started to change tickets with each other to get the fields they wanted. Loi at the People's Committee also suspected, or at least suggested, that maybe not all of the fields had ended up in the lottery either, but were already 'spoken for'. There was also a third explanation, namely that some families of soldiers who had died in the war got priority to pick some extra fields before the allocation started. The same had happened with residential land where some of the men considered war invalids had been allowed to choose plots for their houses prior to other villagers.

My landlady, Lan, revealed that she had been swift changing the tickets she had drawn. Her husband was a teacher and therefore not entitled to any farmland. The agricultural matters were her responsibility, and it was her name that was in the Red Book. She was quite content with her present fields. They were of relatively good quality and all of them situated fairly close to the house. Not all the farmers were equally content. First of all, you had had to find somebody to exchange the fields with, and they had to have fields of comparable size and quality. People who might not have qualified to get land in the first allocation when the lottery was conducted, normally found it difficult to swap their fields, as the quality of these later allocations were rather inferior. Lan had, however, not stopped at field-swapping. She had been courageous enough, as she put it, to ask for some extra manioc fields. These fields had, just like the rice fields, been distributed in a lottery. But as the dry hill land where manioc was grown was less valuable than the rice fields, and also of smaller quantity, the allocation of these fields had, as a consequence, been less strict. Her request had been granted and this was, she argued, mainly because many people never had come up with the idea of actually asking for more manioc fields than they had been allocated.

The cadastral officer had not been in charge of the distribution of the manioc fields, but had dealt mainly with the rice fields. Instead, it was a person who had also been a policeman in the commune and a close neighbour of Lan, who had handled this. Unfortunately I could not inquire about these matters as this man had died in an accident some years ago. His widow told me, however, that he had arranged for other persons to pick the land tickets for his households in order to avoid rumours about cheating. By and large, the villagers seemed to be content with how the land had been distributed and they thought that it had been a rather fair process, even if not perfect. Lan's conduct proved that there were some elasticity in the executing of the allocations, but she assured me that there had been no way that she could have asked for more rice fields than her household was allowed to have. One day, when we were sitting by the stream that ran just behind Lan's house, the former village leader stopped by to chat while Lan pointed out some of her rice fields to me. He joked with her and said that as soon as she saw an unused piece of land, she turned it into a rice field. She had, according to him, the potential of becoming a true 'landlord' (*dia chu*).

Clearing little pieces of unused land was what many of the farmers had done when there was still land available to clear. Even during the collective period, there had been some small pieces of land here and there that were not cultivated by the collective. Here, the villagers had grown things like vegetables and beans for household consumption. Some of the households still held these kinds of fields. They knew that in theory the commune could revoke their little plots, but in practice this was not very likely to happen since the areas were small and the quality was not the best. Today, it was not possible to find any land to clear for growing even the smallest amounts of vegetables.

Many of the households had turned a patch of land into a fishpond, usually fairly small ones, either in their gardens or out in the village. The fish were mainly for household consumption. However, fish thieves passing by might steal the catch before it landed in the owners' dinner bowls. Fishponds were a male concern, except for feeding the fish that was sometimes made by women. Men who had access to large ponds could often be seen standing by the water's edge in the early evenings, gazing at the surface that was sometimes broken by a moving fish. During my stay in the village, the cadastral officer, Han, got to rent the largest fishpond in the village from the commune, and from that point on he was often seen in contemplation by his pond as darkness approached. Before this, he, unlike his farming wife, was rarely seen spending time outdoors. His brother-in-law did most of the work with the pond, even if Han occasionally participated.

Land related discontents

On a couple of points, however, the farmers conveyed united criticism of the land distribution. For instance, some of the people who had been allocated land were not actually supporting themselves as farmers even at the time of allocation. In other cases, people had changed their occupation after the land distribution and were currently possessing land use rights that they did not really need. This irritated the farmers who felt that arable land should be held available to the actual farmers. It was not fair that the farmers, especially young couples, had to rent fields from non-farming persons. The most massive discontent, though, concerned the fact that the families of landholders who passed away could keep the land of the deceased person until the presumptive future allocation while (and this was the real catch) new-born children did not get any land at all. This had the result that households formed after land had been allocated did not get any additional land when the family grew. One woman, for example, had had one child before the allocation and one after, and she thought that in some respect the youngest son ate from the field of the older son. She, and several farmers with her, argued that it would be better to allocate the fields for 20 years – just like the 1993 Land Law did – and also make an update of the situation of each household every three years or so. In this way, land held by deceased persons could be passed on to babies. The vice chairman at the People's Committee held similar thoughts when he argued that farmland should be available for the use of active farmers and if such a procedure should be implemented it could also compensate for the lack of land to feed the farmers' young children. 'To avoid social unrest' as he pointed out 'it is important to ensure the well-being of the farmers.'

The relatively stable land situation in Lang Xanh nevertheless involved elements of insecurity. There were villagers who had lost smaller areas of allocated land, for instance when a road had been widened or the pre-school had been built. They had not received the compensation they felt they were entitled to, and some of them had given up hope of ever having the matter solved. The household of the teacher Luc and his farming wife, for instance, had lost one of the fields where they used to grow sweet potatoes. That field used to be where the volleyball ground – the only appointed public recreational area in the village – now was. Since this was a kind of public service, they thought, they would probably never receive any sort of compensation. Cong, who had been allocated most of his fields at a later stage, due to his absence as a soldier and his wife's former state employment, called his small area of dry fields 'borrowed land'. This kind of land was higher and dryer than the rice fields and therefore he thought that the commune might retract it to use it for residential land, especially as it was close to a road. He felt that the rice fields were more secure, but thought that even residential land, which was allocated without a time limit, could be lost. 'If the state decides to build a road through your house you will lose it', he asserted, 'but then you might get compensation.'

It was made clear to me already at an early stage that some aspects of land were sensitive to talk about. The cadastral officer and the village leader, who had the most insight into the circumstances of land in the village, were both very reticent when it came to matters of land control and distribution. When I started to do interviews with the union leaders in the commune, I often asked if they dealt with any land issues as part of their union activities. This commonly caused the interviewee to assure me that there were no land conflicts in this area, they all got along well and helped each other out whenever it was needed. From their official standpoint, land issues usually meant problems and the union leaders did not want to deal with land from any kind of legal perspective. All of them told me what I was going to hear many times during my year in the village, namely that the land policies were determined by the state and they had to accept and follow these policies.

Some of the villagers were slightly less tongue-tied about the allocations, and about the informal land transactions in which most of them engaged in order to adjust their land holdings according to the changing situations of the households. Land conflicts were, however, a topic they tried to avoid. As a general rule, misdeeds and disputes were not supposed to be revealed to outsiders. The villagers frequently emphasised that there were no (or at least very few) land disputes. I took this as evidence of the sensitivity of the matter. Even so, to my knowledge, it seemed to be correct that there were no big land disputes going on in the village. Nonetheless, there were constantly smaller disagreements going on, especially concerning the size and maintenance of the banks that divided the rice fields, or the drawing of the borders between residential land.

The issue of water for irrigating the rice fields was another cause of quarrels. At one of the village meetings that I attended, a woman complained that when she asked the man in charge of the large water dam in the outskirts of the village to have the water turned on, he had replied that she would first have to go to the village leader and ask him. But when another woman had asked for the water to be let into the area where her field was, he had turned it on immediately. It was agreed that it was important to sort out the water issue and that the water dam was meant to serve everybody, and not only for the man in charge of the dam to rear fish in. The woman, who sat beside me at the meeting, quietly remarked that this matter was brought up every year, but that it had not yet been properly solved. One single person was responsible for the main water supply in the village – a large dam that was owned by the farming cooperative (which still existed but in a different form, a subject I shall return to later) which charged a fee from the farmers twice a year – and he was elected at a village meeting. His job was to turn the water on and off in the areas that contained rice fields and he was supposed to direct the water to and from the fields depending on requirements. This had to be coordinated to a certain degree, and it was important for the farmers to know when they could have water led into their fields in order for them to prepare the fields correctly. Another woman at the meeting explained to me how the water supply for the rice fields was organised and she assured me that water for irrigation was the most common reason for quarrels, more common than land border disagreements.

Ngan and Hieu, two faming women in their forties, on the contrary, held border infringements as the number one reason for disputes. One afternoon in Ngan's house when nobody else was around, the two friends explained that there were plenty of occasions when some greedy person had tried to get some extra land, for example, by planting trees a little too far into the neighbour's garden – even if people on both sides knew where the border actually was supposed to be. The bamboo sticks that I had seen firmly hammered into some of the banks that demarcated the rice fields, had a practical explanation. Ngan and Hieu described their use as markers of the middle of the banks. If somebody felt that the person using the field on the other side was trying to enlarge their own field a bit by reducing the bank, often with the pretence to make the bank look neat and straight, then bamboo poles were put deeply into the bank. It would then be obvious if the neighbour was making the bank smaller and the field a little larger. When describing these land disputes, Ngan and Hieu were both rather embarrassed, and they chose to give examples of quarrels that had taken place some time ago, or they discussed the disagreements in a more general way, without giving specific examples.

The farming cooperative

On the other side of the road from the People's Committee, surrounded on three sides by rice fields, the rather run down buildings of the former farming cooperative were still to be found. These walls were also plastered in yellow, even if the surface was tarnished. The largest part of the space was left vacant for most of the time, but the row facing the road was still in use. This was where the present farming cooperative had its office. Nowadays the membership was voluntary, I learnt in a conversation with the leader of the farming cooperative, a woman in her 50s named Ly, but in fact all of the farming households in the commune were members. The approximately 500 households of the commune had provided 1046 members. There was no annual fee for membership in the cooperative, but there was an introductory fee of 28,000 VND[10] to be paid by each member. If the cooperative made any profit, it was, at the time, used to fund its activities. People in this area were poor, hence the decision not to charge an annual fee, and when a member passed away, the family received 50,000 VND from the cooperative, the leader informed me. The activities of the cooperative did not involve actual

10 100,000 VND ≈ 6,5 USD (April 2003).

farming anymore. It did not have any land at its disposal and instead the focus was on providing services to the farmers, like selling high quality seeds and fertiliser for reasonable prices and giving courses in farming techniques etc. Nowadays, as Ly pointed out, the farmers could buy their seeds and fertilisers wherever they wanted, but to buy them from the cooperative was safer and ensured good quality. Customers who were members of the farming cooperative were given discounts on their purchases, but non-members could buy agricultural products from the cooperative to market prices. Most of the customers were women and there was a slight majority of female members. The farming cooperative also gave some guidelines, about for example when to transplant and harvest rice and it was the duty of the village leader in each of the commune's villages to pass this information on to the farmers at village meetings. Many of the farmers seemed however to ignore the meetings and instead organised their farming by talking to each other whenever they happened to meet.

In the old type of cooperative the leader of the commune's Communist Party was also the leader of the collective farm. During my stay, the leader still needed to be a Party member. Also the agricultural cooperative was under the control of the People's Committee, but unlike the rest of the political and administrative structure, there was no national organisation for agricultural cooperatives. These existed only at communal level, and cooperated with all the unions to spread information about their activities. At the time, the positions as leader, accountant, and treasurer were all held by women.

Lang Xanh had experienced around 30 years of collective farming and the leader of the commune's Communist Party told me that it had all begun with the first cooperative being formed in the province of Thai Nguyen in 1958. Locally, the land situation changed when the cooperatives in this area were formed in 1959. In the course of only a few years, all land became the property of the state and was eventually organised into collective farming cooperatives.

Chi, who was close to her 80s, had experienced the changes in land politics first hand. She took pride in having been a hardworking woman all her life, but after the revolution in 1945, she had been particularly burdened by all the work she had to do. Her husband had been a cadre at the commune office, which meant that he did not participate in the household farming activities, and her parents-in-law had owned several *maus*[11] of land, and this required a lot of labour. On these fields, people had worked as paid labourers. She told me:

> My mother-in-law used to employ poor people. At this time people were very poor and came here to look for work. These people were kind of adoptive children (*con nuoi*) for her, but it resulted in my parents-in-law being regarded as landlords. In 1953 my husband was sent to fight at the front in Dien Bien Phu and I and our first son stayed with my parents-in-law and I had to do everything on my own, like working on all the rice fields. At that time nobody wanted to help a 'landlord' family!

She remembered how land reformation was implemented in the village:

> The state did not take all the fields at one time, but took them gradually and gave them to peasants. Often they took the fields just after they were transplanted! These government policies were not good. It could happen that a landlord who had lost all the fields and had to work as a hired labourer got some fields back, because now they were peasants and not landlords. I have experienced many regimes, from when Vietnam still had an Emperor, and it is always the farmers who have the most difficult lives.

Even so, she apparently felt the need to give an example of something good that the state has achieved, so she concluded: 'The crops are much better now and this is thanks to the new seeds that the state introduced.'

When collective farming was introduced in 1959, it was not only fields that had to be handed over to the collective. Buffalos, cows, ploughs and even hoes had to be turned in. 'All the tools had to be given to the collective' the farmer Thang, who worked in the collective for some time before he became a soldier, remembered, 'and instead of making our living directly out of what we produced, we had to work for points.'

At present, if there was one thing that the villagers seemed to agree on, it was that the land situation had greatly improved since the collective farming had ended. According to the stories I was told, the collective farm had clearly been dysfunctional and the villagers considered things to have improved now that the households were in charge of agriculture. People had more freedom to make their own decisions in farming, organise their work as they pleased, and reap the benefits of their labour

11 1 *mau* = 10 *sao*, i.e. 3600 square meters.

efforts, and this was all for the better. Many still had vivid memories of the constant lack of rice and other foods during the time of collective farming, and the access to necessities like salt and fabric was often inadequate. All the hard work that people were expected to contribute had not been adequately rewarded. 'Now the farmers are allowed to think for themselves', as one man, old enough to clearly remember the time of collective farming in Lang Xanh, put it.

Not only has the political map of the village been redrawn, the administrative maps have also changed as old policies have been replaced with new ones. Kha, a retired teacher more than 80 years old, told me that this commune had been allotted its present borders the same year as the cooperative was set up. The village was named Lang Xanh, but was often referred to as *khu* 10 (division 10) by people who lived in the area. This was a remnant from the farming cooperative, which was divided into subgroups. All the villages in the commune were still referred to as their subgroup number and I only learnt to distinguish them by using the numbers. The subdivisions had altered during the years of collective farming. Division 10 had at one time been divided into two, but then merged into one again. But the name Lang Xanh, I was told, had persisted and went much further back in time than the cooperative. As one woman put it, 'Lang Xanh is in division 10, you can say'. The size of the fields had changed as well, during collective farming. Many villagers remembered that the fields had been made as big as possible to suit the working brigade's collective duties. Nowadays, when each household had been allocated a number of fields to use, the banks that marked the boundaries between the fields were greater in numbers, making each field smaller.

Mai, a farming woman in her mid 30s, married and with two young sons, told me how relations in her family were affected by the turbulent times of land reformation and establishment of collectives. Her grandfather had been a rather wealthy man, owning many fields and employing labourers to work on his land. In 1954, after the French had been defeated, the land reformation was gradually implemented in the north and all the land was taken away from the family, even the rice stored in the house was taken from them. She told that one of the old men in the village had, as a young boy, been taken into the family by Mai's grandfather. The boy had lost his parents at an early age and he had grown up more or less like a brother to her father, who was his parents' only child. When the family lost all their land to the collective it turned out that the then young man, living like a son in the family, was a supporter of the revolution and turned against his 'adoptive father' (*bo nuoi*). Their relation was immediately broken off and he was no longer considered part of the family. But, she continued, when her grandfather eventually died, the rejected man showed up at the funeral and expressed his wish to resume his relation with the family again. As Mai's father did not have any siblings and wanted some more people in the family, he accepted the proposition. The two men, both now in their 70s, lived close to each other, but things never became the same again, she asserted.

The dinh *house*

Another remnant from the country's turbulent history was marked by an absence rather than a presence. When proceeding further into Lang Xanh from the village leader's house and looking across the road from where my host family was living, a rather ordinary residential house could be seen that was the home of a middle-aged couple and their teenage son. If this had been in the 1940s, there would have been a *dinh* house there instead. A *dinh*[12] is a communal house where the altar of a village guardian spirit is placed and where the (often mythological) founder of the village is worshipped and commemorated. This is, or in the case of Lang Xanh, was the place where communal rituals were regularly conducted. When the French troops tried to re-conquer this area around 1947, they often used places such as communal houses for accommodating their soldiers. Also, the communist administration, which existed in opposition to the French, did not approve of the village rituals and ceremonies staged in the *dinh* houses. When land reforms began to be implemented in the 1950s, such spiritual places were normally seized and used for much more mundane purposes. The communist regime employed this seizing and redefinition of sacred land as a significant part of their upheaval against the previous governing and land control (Malarney 2002: 13, 44–46). If the *dinh* houses survived this period, they were often heavily damaged. Gradually, the political climate has become less severe on these matters and some villages have rebuilt or renovated their *dinh* houses (ibid.: 189–207). In Lang Xanh, the *dinh* was totally destroyed and the ground was turned into residential land.

12 See Hickey (1964: 218–21) for a description of the traditional *dinh*.

I was presented with several stories about when and how the *dinh* was torn down. Some held the French responsible, while others put the blame on the communists, or on local resistance activities. The commune's Communist Party leader and the leader of the Veterans' Union in Lang Xanh told me that in the beginning of the collective period, the ground where the *dinh* used to be was used for keeping buffalos. In 1947, the *dinh* was destroyed by locals in order to keep the French from using or wrecking it. Today, the two men said, people know that it was wrong to destroy a thing like that, but at that time, the circumstances were different. By word of mouth, some people still remember that the founder of the village supposedly was a man called Han Van Dong who, legend told, was a great warrior. The Communist Party leader told me how the villagers used to sacrifice a large black pig to honour him. It was said that the pig was only scalded and not cooked, and this was to show what a fast fighter the man was; he did not even have time to wait for the pig to be cooked before he ventured into the next combat. Apparently, the *dinh* had been very big, with a beautiful thick tile floor framed in large wooden beams.

Other villagers told me that the *dinh* was torn down by the communists in 1954, as a result of the 'land revolution', and turned into a shed where buffalos were kept. Nowadays, as mentioned before, the destruction is disapproved of and some villagers wanted a new *dinh* house. However, the cadres at the People's Committee were mostly too young to remember the *dinh* and, according to some, did not care about it. One of our neighbours, a woman in her 50s, told me that since the village lost its *dinh*, many villagers had met bad luck, and anyone who lives on the land where the *dinh* used to be will meet difficulties, according to the woman. She went on to explain that the present male inhabitant was the third generation of his family to live in a house on this plot and it had brought bad luck to all of them. The husband's mother did not get along with her mother-in-law, his wife did not get on with his mother and now he and his wife are getting divorced. Moreover, most villagers are hardworking but nobody gets rich here, she pointed out and continued, 'according to uncle Ho[13], good knowledge in the hands of the wrong people can be disastrous and this is what happened when the

13 Ho Chi Minh was, and still is, sometimes referred to as Uncle Ho (*Bach Ho*). This can be said to signify the combination of respect and closeness that was characteristic for his leadership (Marr 1981: 100).

dinh was torn down. The general idea of land reformation was good, but to destroy the *dinh* was very wrong'.

My landlord's mother maintained that the *dinh* house was destroyed by the French, along with a pagoda (*chua*) that used to stand on a hill in the outskirts of the village. When she moved here from Phu Tho town in 1945, the *dinh* was still here. For some time, people talked about rebuilding the *dinh*, but it never happened and nowadays nobody mentions it anymore, she said. The actual circumstances of the destruction of the *dinh* and the pagoda were obscured by the disparate stories presented to me; each story coloured by different ideological positions and individual apprehensions of the course of events.

The pagoda

Close to the crossroad, where you turn right from the main road to enter the village, was a rather low hill with some thin trees surrounding a pagoda. Even though it was situated on the outskirts of Lang Xanh, the pagoda served at least the entire commune. First and foremost, this was a popular destination for the older women[14] in the area and at spiritual feasts and celebrations the pagoda became quite crowded.

As mentioned above, the old pagoda had been destroyed together with the *dinh*, but unlike the communal house, the pagoda was now rebuilt. This had been possible mainly thanks to the engagement of and pressure from the older members of the commune, who also had initiated fund-raising in order to get enough money to offer to the People's Committee. The idea had been to make it difficult for the cadres to refuse, using budget deficit as an excuse. People's wish to rebuild the pagoda was finally accepted and the project was even allowed a grant from the commune. The new building was fully completed in late spring when I was there and the celebrations were anticipated with great expectations. Well in advance, my landlord's mother started to urge me to be present

14 Hy Van Luong notes that in the village of Son-Duong, as in many other Vietnamese villages, there are mainly women who are engaged in the Buddhist Association (2010: 231). This holds true for the cities as well. Alexandre Soucy writes that between 80 and 90 per cent of the attendants in Buddhist pagodas in Hanoi are women, nevertheless, all the major positions within the structures of the Buddhist organisation are held by men (Soucy 2012: 3). Women often carry the responsibility of caring for the spiritual needs and welfare of the household members (more on this in chapter 3), and this can be said to form part of the ideas about caring, trusting, and delicate femininity (Soucy 2012).

and to bring my camera to commemorate the occasion. Together with other important guests, a prominent Buddhist monk was invited and a large crowd turned up to join the festivities, make contributions, say prayers, and give offerings to the divinities and the organisation of the pagoda and, finally, bring some blessed colourful rice-flour tops home as auspicious tokens to the ancestors and family members.

My landlord's mother told me that most people in Lang Xanh found the pagoda to be more important than the *dinh*, and that there was no other pagoda in the vicinity. She, like many of the old women and a few of the men, was a member of the local Buddhist Association. For them, the pagoda was the centre of their activities and an important social and spiritual institution. So even if the previous pagoda, according to the old man Chanh, had been larger and more beautiful – with a bell that, when it was sounded, spread a delightful chime in the wind. At least the pagoda had been replaced.

The culture house (**nha van hoa**)

Lang Xanh now had a 'culture house' but, as the mother of my landlord explained, this was a state concern and did not contain any altar or divine figures, and there were no connection whatsoever with spirits, mythological entities, or other unearthly beings. This house was situated in the outskirt of the village and built just recently, when Lang Xanh was appointed a 'cultural village'. Old Cuong, a former cadre, informed me that this appointment meant that the Lang Xanh inhabitants needed to be better than other villages in every way, in terms of economy, living standard, hygiene etc. But most important of all, they needed to avoid disputes and show solidarity with each other, work together and help each other, and get along well.

The village meetings were held in the cultural house and the village leader sometimes reminded the villagers of Lang Xanh's designation as a 'cultural village' in order to urge the households to pay taxes and fees, and to try to discourage the two couples who were in the process of getting divorced to reconsider this decision, as this was not in line with the criteria for a cultural village.

May was the first anniversary of Lang Xanh being a cultural village and this was celebrated by a number of song and dance acts performed by men and women, young and old and organised mainly by the Veteran's Union, the Old People's Union, the Women's Union, and the

Figure 2: Old women on their way to the pagoda

Youth Union. The wife of Loi was the presenter, while her husband, who was also the vice-chairman of the People's Committee, was at home with their two small boys. This was not overly remarkable as women generally showed more engagement in these kinds of events.

Places for socialising

Neither the cultural house, nor the People's Committee were places where the villagers, regardless of age, came to socialise with each other. These were public buildings, but they were mainly used for official meetings and organised events. Apart from the volleyball ground, there were really no designated public areas where the villagers gathered. For old people, particularly old women, the pagoda was an important place for meeting and socialising with each other. As for the rest of the villagers, it was out on the roads and paths – usually on their way to or from the fields or some other work engagement, or going to the markets – they met with friends, relatives, and neighbours. Quite a lot of socialising also took place in people's homes. Especially in the evenings, after dinner, people dropped in on each other to have some tea and a chat. This kind of leisure socialising was highly gendered. In general, men spent much more time visiting their friends and eating and drinking together. This usually happened in someone's home, where the women of the household made sure that the men had the food and drink they needed, while they themselves retreated to the back parts of the house. Men also went out on their own more often than women did. If women went out to visit someone on a less busy evening, they usually went in couples or in small groups. As we shall see later on, women found time to chat and socialise not least while they were working together in the fields. In the evenings, women spent most of the time doing laundry, feeding the pigs and performing a myriad of other household chores. The following chapters will explore how primarily women's work oriented them in certain directions and how this generated gendered understandings of access to both residential and arable land, as well as gendered divisions of labour and contributions to a sense of village community.

CHAPTER 3

Continuity and change – gender and land access

This chapter explores how certain 'female' and 'male' capacities were closely intertwined with perceptions of, and access to, two particular types of land: residential land and farmland. I will show that, in Lang Xanh, women and men gained access to farmland and to residential land on the basis of different criteria. They were, one could say, oriented towards these two types of land in different ways. The histories, meanings, and definitions of farmland and residential land respectively differed markedly in the village, and this held significant implications for women and men's preconditions and opportunities to gain access to these types of land.

As for residential land, the plots were inseparable from the actual houses, as well as from their inhabitants. The connections and dynamics between this type of land and the dwellers were marked by a sense of continuity in access practices, and gender played a decisive role in how access was acquired. This contrasted with the rice field, to which the Lang Xanh residents had a much more temporary and uncertain relation, as previously explained, and where notions of gendered access had gone through radical changes as a consequence of the drastic land regimes brought on by the 1945 revolution.

This chapter will mainly focus on how understandings of gender-specific abilities interacted with perceptions of residential land and farmland respectively, and informed the ways in which these land categories were available to men and women.

Dwelling, building, and gender

Heidegger argues that dwelling is the mode whereby humans exist on this earth. By drawing on linguistics, Heidegger traces this human mode back to *buan*, the Old English and High German word for dwelling. He shows that, etymologically, 'being' and 'dwelling' were (in German) originally the same thing; rhetorically he asks, what then does *ich bin* (I am) actually mean? 'The old word *bauen*, to which the *bin* belongs, answers: *ich bin, du bist* mean: I dwell, you dwell.' (Heidegger 1971: 145) Eventually, the word *buan* took on the meaning of building (*bauen* in German), and thereby the active and generative original meaning, in which this word constituted the very essence of how humans took up their existence in the environment, was lost and turned into a more narrow and less dynamic meaning. However, according to Heidegger, the concept of building still contains traces of its original meaning, taken to mean both constructing and cultivating, it is indicative of how humans inhabit the world, i.e. how we dwell on earth. In other words, building is intrinsic to the human condition, and it is by building that the earth is gathered into particular locations and sites. A building, for example, is not erected in a predetermined location; on the contrary, he claims, it is the building that makes the location emerge by drawing together the premises of the human life-world into a meaningful locality (ibid.: 144–45, 150–52). Although Heidegger quickly abandons the cultivating aspect of human dwelling to pursue the line of building – something that we will return to – I will employ this perspective on cultivated land as well. Building a house or cultivating rice on a plot of land are two different acts, but both acts are constitutive for the person who performs them, or – equally important – *not* performs them. This means that fields, too, can be taken to have the capacity of gathering together conditions for human existence into locations of mutually constituting meaning, i.e. places where people and the world come into meaningful being.

If building and cultivation are constitutive to our situation as human beings, it is highly significant for us as dwellers if and how we can engage in these two acts. If building a house on a piece of land, for example, is a major constituting act and one which is predominantly perceived as being a male capacity, the practice of house building marks a difference in ability between men and women. Hence, when a man builds a house on a piece of land, this act constitutes him as a man who builds. And the place becomes a site with a house that is associated with a particular male builder. This sets a building dweller apart from a dweller who mainly cultivates the land. However, one important condition for how we dwell in places, which neither Heidegger nor Ingold take into account, is the gender of the dweller. A gendered division of labour, as this book explores, make acts of building and cultivating become not only ways of being in the world (dwelling), but *gendered* ways of existing. In Lang Xanh, men and women's involvement with residential land and farmland – i.e. building and cultivating – was highly informed by understandings of gender capacities. By doing different things in different places, men and women were oriented towards the world in gender-specific ways.

It is important to point out that even if places and people are mutually generated, individuals are born into already existing relations between people as well as between the dwellers and their environment. The engagement of previous generations in the environment sets the preconditions for later generations, who experience the presence of their predecessors in what the ancestors have left from their previous dwelling in the world (Ingold 2000: 149, 186, 189). The landscape, Ingold insists, bridges generations of dwellers by being a kind of narrator of the life of former dwellers. By not conforming to the idea of a linear time, where every moment is like a bead on the string of time that disappears into the past, the idea of dwelling allows people to experience their ancestors as an embedded part of life. Ancestors' earlier presence is interwoven into the texture of land and life, which makes the beginning and end of each generation less distinct (ibid.: 189). As we will see, such a view complies well with the villagers' idea of ancestors and their relation to residential land. In Lang Xanh, the regular practices of ancestor worship and reverence for deities, bestowed houses and land, as well as women and men, with constitutive qualities and, thus, positioned all of them in relations to each other in gender-specific ways.

The formation and maintenance of the inhabitants' social and spatial relations depend greatly on the close connection between family homes and residential land, and ancestors and the other world (*the gioi khac*). A house is not an isolated unit but stretches out into a broader societal context, thus linking the inhabitants to for instance state politics and economy (Carsten 2004: 37). Anthropological studies have paid considerable attention to the extensive importance of the house as a link, or a key, to cultural understanding (e.g. Bourdieu 1990; Carsten

2004; Carsten and Hugh-Jones 1995; Gudeman and Rivera 1990; Lévi-Strauss 1983). The site that contains the house, on the other hand, has received comparatively little consideration. However, Carsten and Hugh-Jones remark that 'the space that surrounds a house is also an extension of the personal space of its occupants' (Carsten and Hugh-Jones 1995: 3). As we will see, in the case of Lang Xanh, the piece of land that harboured the house bore far more significance than this, and was therefore of utmost importance for the people who dwelled there. Moreover, according to Ingold and to Heidegger, a house and its ground are inseparable in the sense that they form a prerequisite of, and give meaning to, each other. This corresponds well with the house-related practices of the Lang Xanh peasants. As anthropological studies have shown, from a gender perspective, a house is an integral part in the making, naturalisation, and sometimes contestation of, gender relations (e.g. Bourdieu 1990; Carsten 2004: 50; Carsten and Hugh-Jones 1995: 21). From my fieldwork in Lang Xanh, it was apparent that this holds true also for residential land.

Ancestors, residential land, and gender

For most families in Lang Xanh, the relation with the ancestors was an integral part of life. This was made evident not least by the fact that virtually all residential houses had at least one altar devoted to forefathers, principally – but not always exclusively – for the patrilineal ancestors of the male family head. Ancestor worship was often spoken of as a duty, in the sense that it was the responsibility of the living members of the family to honour and respect the former generations, not only while they were alive, but equally important, when they had moved on to the other world (*the gioi khac*). This transference did not mean that the ancestors ceased to exist; they continued to be around, together with other kinds of spiritual beings, but now in the other world. Thus, the relation continued in the form of worshipping at altars in the home.

Place and ancestors

The significance of ancestors' connection to certain places was made unmistakably clear to me by Sinh, a married woman in her 60s who, after having enquired about the Swedish way of worshipping ancestors, learnt that Swedish people commonly did not feed or worship their ancestors at altars in their homes. She looked at me, clearly alarmed, and concluded that in that case Sweden must be full of hungry and wandering ghosts. Here in Lang Xanh, regular ancestor worshipping and anniversaries, preferably conducted in the house of the eldest son, made sure that ancestors were nourished and equipped with the things they needed in their new shape and place of existence [1]. Ancestors needed a particular place to come to in order to be fed and cared for, and the home-altars provided such specific places.

In the household where I lived, it was my landlord's mother who usually performed the twice monthly regular worshipping of the ancestors of the family's patrilineage. The altar was placed on the wall above a large chest of drawers and held three incense bowls, an oil lamp, and a vase with some decorative paper flowers. When the worshipping was due, the old woman usually dusted the altar and the piece of furniture, put fresh incense sticks in the bowls and started with arranging items of food that she had bought fresh from the market. She arranged the food on small plates made of plastic or china and put some of these on the altar and the rest on the chest of drawers beneath. The actual worshipping commenced when she lit the incense sticks, bowed before the altar and began to invite the relevant deities and the ancestors to come and eat from the offerings. The burning incense sticks was the link that opened up the connection to the other world and made it possible to invite otherworldly beings to the altar. When the sticks had burnt out, the ancestors and deities had finished their meal and the connection was broken.

Common food items to offer were cakes made of sticky rice, fresh fruit, nicely packed crackers or snacks, betel nuts accompanied with lime paste and a special kind of green leaves, and small cups of rice wine. Some cigarettes were usually included as well. The fruit varied according to season, and when the family had a visitor who brought something delicious to eat, which was a common practice, this too was offered to the ancestors for them to enjoy. Cooked food was normally offered on major events, like weddings, funerals, anniversaries, or holidays. Then

1 Objects such as clothes, money and gold was passed to the other world by burning votive papers and paper items symbolising these offerings. These kinds of offerings were mainly made at larger events such as New Year (*Tet*) and death anniversaries. Usually the burning took place outside the house, in the courtyard.

Figure 3: The ancestor altar in my host family's house

the ancestors were presented with samples of the most delicate pieces of the newly prepared food and when they had had their share, i.e. when the incense had gone out, the food was removed from the altar and could be consumed by members of the family who were living in this world.

If the ancestors were neglected, dissatisfied, or lacked an altar to return to regularly for their needs, they could turn into wandering ghosts that roamed around restlessly and troubled their descendants or other living persons (see also Gammeltoft 1999: 70; Kwon 2008: 17, 20; Malarney 2002: 117). Most susceptible to become a wandering ghost was a person who did not get a proper burial and was not worshipped by any relative from his or her patrilineage[2]. When close relatives – preferably children and/or a spouse – performed the funeral rituals, the soul of the diseased would gradually be transformed into a 'divine soul' (*linh hon*) who could be called back to the house altar to receive the offerings and, in return, offer the living family members its protection and support in virtually any endeavour they embarked upon (Chanh Cong Phan 1993: 172–74, 177–82, Rydström 2003: 85–88). However, if the funeral rites for some

2 See Malarney (2002) for an extensive study of contemporary funerals and how such practices have been influenced by political changes.

reason could not be performed, the soul of the dead person would leave the body, but instead of transforming into a divine soul, the soul would become a 'ghost soul' (*hon ma*). If at this stage no rituals were performed to call the soul back to the house where the family of the diseased lived, the ghost soul would turn into a proper ghost (*ma*) who usually drifted around restlessly at the place of death (Chanh Cong Phan 1993: 174). As Kwon notes, a ghost is not at home in this or the 'other world', they 'are the *nguoi ngoai* – their [the Vietnamese] term for strangers or outsiders – in the world of the dead' (Kwon 2008: 20). Consequently, many villagers told me that one of the most awful things that could happen was to die in a foreign place where no family members could see to their safe transformation into revered ancestors[3]. If this should happen, their souls would wander uneasily for eternity. The ideal end to a person's life was to be buried in one's home village and then be worshipped by a son at the house where he or she used to live. Clearly, this was not always possible, but it was an ideal from which there should be as few deviations as possible. From this perspective, men and women were equally eager to have a son, whose house could provide a safe place for them as ancestors to return to for their needs.

Beliefs concerned with ancestors' ties to the home place were perhaps strengthened by the custom of caring for the ancestors' tombs that traditionally had been placed in the fields or gardens belonging to the family. However, nowadays the ancestors' tombs usually were gathered in a site designated by the Peoples' Committee in the commune, as the previous chapter mentioned. In his study of the changes of the northern Vietnamese family, Pham Van Bich describes how, in pre-revolutionary Vietnam, the sons' duties towards their forefathers made them reluctant to leave their home villages even when times were hard. The sense of rootlessness was often severe in a place where the land belonged to other families. A male newcomer was commonly perceived as a total stranger, due to his lack of ancestral connection to the place, and it could take generations before a family was considered 'insiders' and allowed to enjoy the same privileges

3 In *Ghosts of War in Vietnam,* (2008), Heonik Kwon writes about how the war between the USA and Vietnam resulted in great numbers of wandering ghosts. The men, women, and children (Vietnamese as well as foreigners) who died a 'bad death', a 'death in the street' (*chet duong*), i.e. a death away from home (*chet nha*), on Vietnamese ground became ghosts who had to wander around lacking a place to settle in.

as other villagers (Pham Van Bich 1999: 20, 27). Hy Van Luong writes that the terms *noi tich* (insider) and *ngoai tich* (outsider) were used in northern and central Vietnamese communities to clearly and severely differentiate between those who belonged to a place or a kin group. If a village had communal land to distribute, it could take three to four generations before a male member of a newly arrived family could have his status transferred from 'outsider' to 'insider' and receive a share of this land (Luong 1992: 229–30). Only those men who appeared on the village member lists could partake in the allocation of communally held land, and this contributed to their disinclination to move away from the home village. According to Wiegersma, women's names never appeared on the lists (Wiegersma 1988: 54–55). The bond between men, home place, and land was strong.

Some of this place-based anxiety prevailed in contemporary Lang Xanh. Men were more often than women expected to maintain close relations with their native family and village and, preferably, one son was obliged to inherit the house and care for the parents in this and in the other world. A son's house, ideally the home of the eldest son, served as a place-bound anchor for the souls of the ancestors to return to. Without such a place to serve as a bridge between this and the other world, the ancestors could never find their otherworldly places either, and therefore be forced to take the shape of wandering ghosts (Chanh Cong Phan 1993: 187).

Place, residential land, and deities

Besides the ancestors, other spiritual beings inhabited residential land and influenced people's lives and aspirations. Reminiscent of how the ancestors were honoured, these deities and spirits were addressed through home-based worshipping.

The Land God (*tho dia* or *tho cong*) was involved in all spiritual and otherworldly dealings that were conducted at home (as opposed to the pagoda), and he was the first divinity that people called on when they wanted to contact their forefathers. It was through his guidance that the ancestors could find their way to the altar and to their living descendants. It was commonly believed that one Land God was in charge of an area roughly corresponding to each of the five hamlets in the village[4]. When

4 Nguyen Tu Chi writes that in the traditional villages of northern Vietnam, each hamlet had an altar, usually at the entrance, devoted to their *tho than*, i.e. Spirit of the Soil (Nguyen Tu Chi 1993: 63).

a household intended to worship the ancestors, the session would thus start with an invitation to, for example, the Land God of the Manioc Tree Hamlet. The clearly place-bound characteristics of the Land God made him a link between the ancestors and the persons inhabiting a particular piece of land within his 'jurisdiction'. As we will see below, when a new house was about to be built, a ritual dedicated to the Land God was performed by a senior male person on the very first day, before the work had commenced. This was to inform the deity and ask for his blessings.

I first heard of the Land God when Loan emerged from a house opposite from where we had been sitting in the shade of the old agricultural collective building, chatting to the woman running the shop on the other side of the road. Loan was a woman approaching her 60s, who communicated with beings in the other world. She was not exactly a spirit medium (*ba dong*) who incarnated deities, but she believed – and so did some of the villagers – that she had the ability to connect with the spiritual world by praying and making offerings to the deities (a kind of *thay cung*). She lived by herself, with her married son living in a house in her garden, and her unofficial third 'husband' only a couple of minutes' walk from her home. Her house had two big altars where hundreds of incense bowls were collecting a fair amount of dust. These bowls had been placed there by local women who had come to see her in some pressing matter, typically related to some kind of marital or family trouble. Now she was on her way to one of my landlord's younger brothers' house to sort out a problem with their small son, and she asked if Phuong (my assistant) and I wanted to come with her. This was a busy day for her, she explained; she had performed a ritual devoted to the Land God, which involved burning a black paper horse[5], in the house she just left. It was not clear what the problem had been about, but the trouble with the little boy in the family she was heading for was obvious; he cried a lot and did not sleep well and his mother had asked her to come and see what could be done.

When we came to the house where the boy lived, Loan was already sure about what to do. The mother of the boy had been instructed in ad-

5 A black paper horse was a common offering for the Land God on special occasions. The only explanation I heard for this was that the Land God needed to cover quite large areas and a horse was an useful item for this purpose.

vance about things to buy for the ritual that was going to be performed and as soon as we arrived, the two women began to prepare trays of food, paper bundles containing small notes with the names of the recipients written on them, and paper items, such as imitations of bank notes, paper 'gold', and small paper clothes, to be burnt as offerings. As usual the ritual commenced with lighting incense sticks to establish connection with the other world, and Loan murmured all the invitations as well as the purpose of the ritual. To check if all the offerings had been accepted by the intended recipients, Loan tossed two old copper coins on a plate, once in front of the altar and once outside in the yard. If both coins showed the 'lucky' face up, the offerings had been accepted and the recipients were pleased. If not, it meant that some deity or ancestor was discontented with the offerings, or felt overlooked and neglected. At first the coins kept showing the malevolent face; some recipient was 'just laughing and turning the back' at the offerings, Loan explained. She had to toss the coins quite a few times before they finally showed the benevolent face (see also Malarney 2002: 100). It turned out that the problems were caused by an unsatisfied ancestor of the boy's father and, after some discussions between the two women, they agreed that it was an uncle of the boy's paternal grandfather. This man had died at quite an early age, before he had any children of his own, and therefore there might not be any family who worshipped him in particular, like a son would do. If this family started to explicitly invite this man when they worshipped, then the boy would probably stop crying and be calmer at night.

During the whole ritual, the mother of the boy was very attentive and when Loan had finished the offerings in a successful way, she asked for some specific advice on how to worship in the future in order to avoid further problems. Loan explained to the young woman that every time she was worshipping, she should start by stating the day's date and the full name of her husband, who was the household's head and whose ancestors they worshipped. She should then proceed to invite the Land God and other deities, like the Kitchen God for example, thereafter giving the names of the particular ancestors they wanted to worship before inviting her husband's ancestors in general. This was the normal procedure for any ancestor worshipping.

Figure 4: Ritual offerings

Except from explicitly mentioning the name of this specific ancestor, the woman was advised to set up a small temple (*mieu*)[6] in her courtyard. Like most people in the village, Loan knew that the house of this family was built on ground where a temple once had been situated. Such ground was very potent, due to the demons and ghosts that tended to live there, or at least spend some time there before they moved on. Setting up a small temple would be an effective way to keep such otherworldly creatures in a benevolent mood and prevent them from entering the house and instead stay in the courtyard, where they could inflict less harm on the family. The young woman listened carefully to all the advice. She explained that she was quite inexperienced as far as worshipping was concerned. In her childhood home her mother had done all the worshipping, and therefore she had never acquired any deep knowledge about these matters. Now, as a married woman, she found

6 The temple she was referring to was a small concrete structure built on a pole. These temple where usually only big enough to contain a couple of bowls of food and some incense.

herself carrying this responsibility on behalf of her husband's ancestors since he, like many men in the village, was reluctant to perform any form of worshipping. Loan's advice about erecting a temple in the courtyard made her hesitant, though. In fear of appearing superstitious (*me tin*), she did not want any passers-by to see that her family engaged in more worshipping than what was normally required. When I met her again later, she told me that they had included the ancestor in the worshipping, but they had not taken any precautions against the ghosts. The boy had, however, become calmer.

The example with the crying boy points to the link between the comfort of the living and the contentment of the paternal ancestors. It also indicates that the land people dwelled on was an integral part of this relationship. Plots of land came with a history and contained otherworldly beings. The acts and events that had taken place on the piece of land that people inhabited became an intrinsic and lingering power that affected the dwellers. The consequences of this were basically dependent on how the male head of the household and his patrilineal ancestors formed a relationship with the deities who inhabited the site. In lucky circumstances, past occurrences did not have any unwanted bearings on the current inhabitants, and a good fit between a piece of residential land and the head of a household could bring good fortune to a family. In less lucky cases, accidental or brutal deaths, ill-fated previous dwellers, or other unfortunate events that had occurred on a piece of land, could cause all kinds of bad luck by the restless spirits who, usually due to lack of a proper place to dwell and people to worship them, had turned into wandering ghosts.

The family living opposite the household where I stayed had their house built on a piece of ground of the old *dinh* (a ritual communal house, see chapter 2). Land that had once held a *dinh* or temple was perceived as unusually prone to ghost infestation because many souls without a home had been gathered there to find some peace and nourishment in the offerings at the altars in these buildings. When the sacral building was torn down, the souls were deprived of their nourishment and dwelling place. As mentioned in the previous chapter, the family who lived where the *dinh* used to be had met several misfortunes. During my stay in the village, the husband and wife decided to get divorced, and they subsequently sold the house to another family in the village.

The new family waited several months after they had bought the house before they moved in; in fact the year of the Goat turned into the year of the Monkey before it happened [7]. The reason for this was that the age – deciding his zodiac sign and, thereby, his 'fate' – of the husband indicated that this would be best for them.

Gender division of labour and place

Just like in my host family, it was commonly the eldest woman of the household who performed the regular ancestor worshipping that took place at least on the 1st and 15th of each lunar month. Some men conducted the rituals themselves, but in most of the village households, a woman, normally the wife or the mother of the male head, carried out this duty. However, it was, virtually without exception, women who bought and organised the items to be placed on the altars. They cooked the different dishes and arranged the fruits, sweets, and drinks that were offered to the ancestors. The men's obligation to keep the ancestral line unbroken did, in fact, require more labour from their wives' than from themselves. The heaviest burden lay on a woman who married an eldest son, since the death anniversaries[8] usually were held in his house. Special events, such as *Tet* (the Vietnamese New Year) and death anniversaries, required a male member of the patrilineage to perform the rituals, but also compelled women to do a lot of cooking for the participants.

Since old women in Lang Xanh were the most frequent worshippers of ancestors and deities, both in the homes and at the pagoda, they usu-

7 The 12 zodiacal signs are: rat, buffalo, tiger, cat, dragon, snake, horse, goat, monkey, rooster, dog, and pig (Hickey 1964: 75). Each animal corresponds to one lunar calendar year. The year 2003, for example, when most of this study was conducted, was the year of the goat.

8 After the first year of the death, when several ceremonies are held in order to assist the soul on its journey into the other world, an annual ceremony is held on the day when the ancestor passed away. This ceremony is ideally held at the home of the eldest son, and normally involves offerings of food, snacks, fruit, alcohol, and the burning of voitve paper items. This is an occasion for the children of the diseased to show their filial piety and the siblings usually contribute to the costs of the ceremony according to their means. The death anniversaries often involve many relatives, and also friends and neighbours, who all share a meal in connection with the offerings. See e.g. Malarney (2002: 138–44) for more comprehensive information about death anniversaries.

ally possessed more experience than other household members in this realm. Elderly women often saw it as one of their main responsibilities to make sure that their families were well taken care of on the ancestral and spiritual level. Prompted by the lunar calendar, which stated all the main worshipping days, the women made sure that the ancestors and the deities were attended to. However, as a consequence of decades of political opposition against spiritual dealings, there was a general fear of appearing superstitious (*me tin*). Too much involvement in ancestral and otherworldly matters could cause rumours to spread around the neighbourhood. One farming woman in her 50s who regularly consulted fortune-tellers and spirit mediums had been given the advice to build a small temple (*mieu*) in her courtyard, the kind that the mother of the crying boy had desisted from getting in fear of rumours of superstition. This woman had taken the advice and her household was the only one in the village to have such a temple. Her husband was not particularly pleased with the arrangement. She knew that some of the villagers thought that she was too absorbed by the otherworldly and the guidance given by fortune-tellers, but she thought that the advantages outweighed the disadvantages. She was, for instance, convinced that the future opportunities of her two teenaged children benefitted from her offerings. Since her house was on the outskirts of the village, the temple was not visible to everybody.

The anxiety to appear superstitious was more common among men than among women, as we saw in the example with the crying boy, whose father largely refrained from practicing worship. Few men engaged much in spiritual activities, and they did not seem entirely at ease when it came to dealing with deities and ancestors. A number of the village women told me that their husbands found performing ancestor worship either slightly embarrassing or somewhat excessive. This quite common male reluctance did not mean that men found the ancestors unimportant, but they did not want to engage much in the everyday practical worship and, thus, left most of the actual performance of rituals to their wives or mothers.

Malarney notes that age, gender, and political involvement are factors affecting Vietnamese people's attitudes towards the spiritual world. He observed that men, especially men growing up in the 1950s and 1960s, were generally more averse than women to involvements with otherworldly matters. This was even more valid regarding those men who had been connected to the party or the army, where so called 'superstitious thinking' was strongly condemned (Malarney 2002: 102). As a general rule, Malarney argues, men have been more engaged in social and political matters, while women have been responsible for the everyday care of the family[9]. This, according to Malarney, has contributed to women's greater engagement in ritual practices as they, apart from the more mundane work, also turned their hopes to the ancestors and forces of the other world for the well-being of their families[10]. The general distance between women and party politics has enabled female family members to maintain the skills as well as their beliefs in worshipping practices, while some men openly despised such procedures and claimed them to be particularly female (ibid.: 102–3).

My landlord was a good example of a man who claimed to disdain otherworldly interactions. For instance, he eagerly argued that the direction of a house was only important from an environmental point of view – meaning that a house should not be built in a fashion where it became too hot or too cold and so on. He believed that the placement of the house did not affect the prosperity of the dwellers due to implications on the other world, as the tradition of geomancy dictated. He did not conceal his opinion about Loan, the woman who was frequently engaged by the villagers in otherworldly matters. Her skills as a fortune-teller and mediator with the other world did not impress him much. 'How come she does not have any money then?' he asked in his typical rhetorical way. 'If she has so much knowledge about these things, why doesn't she win the National Lottery?' Nonetheless, with two adult women in his

9 In the case of traditional Vietnamese village society, principally before 1945, Neil L. Jamieson argues that in contrast to the hierarchical and more competitive male dominated practices of for instance ancestor and village guardian spirit cults, there was a more feminine side centred around Buddhism and pagoda activities. In this context, women, often the elderly, tried to attract support for their families in their particular aspirations by worshipping different spirits (Jamieson 1993: 36–37).

10 A general upsurge of female participation in religious and festival affairs has been noticed. In pre-revolutionary Vietnam, women were not allowed to participate in the communal village rituals, something that they frequently do nowadays, as the occurrence of such festivals has increased after a period of marked decline due to regime policies (see e.g. Malarney 2002: 189–207, Luong 1993; Kleinen 1999: 165).

household (his mother and his wife), his family did not lack otherworldly protection. His mother usually took care of the ancestor worship at the home altar and was also a frequent visitor to the pagoda, and his wife went to see Loan and other fortune-tellers on a regular basis regarding various family matters. In this way, his family was well taken care of as far as spiritual matters were concerned, and he could maintain a slightly ridiculing attitude to what was, in his opinion, an exaggerated trust in unworldly matters. My landlord was a schoolteacher, and as a government employee he was more hard-pressed than other villagers to show restraint in matters concerning the otherworld. Furthermore, he had been a young man in the aftermaths of the revolution and he had, like so many men in the village, served in the army for periods of time. His wife was a farmer and did not have any official or state appointments. This placed her further away from the critical public, or official, eye.

Both male and female ancestors were honoured with regular worshipping and celebrations on their death anniversaries, but these practices were distinctly performed within the realm of a patrilineal system of kinship. For a woman, this meant that if she died after she had married, it was the obligation of her husbands' patrilineal family to worship her. If a woman died before marriage, her natal family would commemorate her as part of her father's family group. Thus, when women engaged in ancestor worship they did it either as part of their fathers' or their husbands' patrilineages. This was illustrated quite vividly by Sinh, who worried about the placelessness of Swedish ancestors. She thought that if she tried to invite her dead mother from the other world while worshipping her husband's ancestors at their home-altar, these forefathers would not let the spirit of her mother in. If Sinh should try to offer her some food out in the courtyard, this would not work either. Even entering the courtyard would be hazardous for the dead woman's spirit, as she would be chased away by the ancestors of Sinh's husband. This was a very lucid example of the difference between 'inside' and 'outside' lineages[11] since it was spatially manifested by literally keeping Sinh's ancestors out. The

11 In Vietnamese, relatives from a persons' maternal side are called 'outside' (*ngoai*) relatives, while paternal relatives are referred to as 'inside' (*noi*). Thus a paternal grandmother is called *ba noi* and a maternal grandmother is called *ba ngoai*. See Rydström (2003: 4, 37–38) for a more detailed explanation regarding the significance of the notion of 'inside' and 'outside' lineages for the understanding of male and female capacities.

land that contained the house where she lived with her husband was clearly delineated as a place where her husband's ancestors, but not hers, could enter. Throughout her married life, the worshipping that Sinh had conducted regularly in her home was directed towards her husband's family line.

Women's transience and impermanence

It was a female capacity to be integrated into another lineage (apart from the woman's paternal one) and also a necessity in order to preserve the husband's patrilineage. Through marriage, Vietnamese women transferred from one lineage group to another – and from one patrilineally rooted house to another – even if the transfer was not total. Women generally retained their ties to their natal families and were expected to participate in the death anniversaries of close family members (Luong 1992: 61; Pham Van Bich 1999: 26). Men, on the contrary, could never become part of their wives' patrilineages, as they were perpetually connected to their own father's family line. Women's lineage mobility placed them in rather more impermanent, or transitory, positions vis-à-vis the patrilineal family, compared to the unending lineage affiliation of men. A woman was meant to leave not only her natal patrilineage, but also her childhood home and move into a house that, through patrilineal ancestor worship, was connected to her husband's family. If she failed to do this, this might have implications both for her and for the house she lived in, as the following example will show.

The single mother Ngoc was a woman in her 50s with a teenage daughter who lived in an old but pleasant house,[12] comfortably situated at the centre of Lang Xanh. She had four sisters and one brother, all married and living in the village, but Ngoc, who remained unmarried, was the one who lived in their late parents' house. According to her sister-in-law, Hieu, the reason why Ngoc still lived in the house was that she was a rather headstrong woman. When Hieu married Ngoc's brother, the father had still been alive. She had moved in with her husband, her

12 As the only house in this village, this home sported a beautiful tile floor. The floor was not complete anymore and other, more common, tiles had replaced the broken ones. According to stories told in the village, the tiles came from French colonial buildings in a small town not far away. When people rebelled against the French in the late 1940s, they broke up the floors in some of the houses, and took the tiles, which were then spread in the area.

sister-in-law and father-in-law in the house where Ngoc now lived alone with her daughter. But, Hieu assured me, it had not been easy to live under the same roof as Ngoc. Eventually Hieu and her husband had found a house of their own on the outskirts of the village, where they now lived with two sons and one daughter. At first, her father-in-law had moved with them, but towards the end of his life, he had gone back to live with Ngoc in the house that he, as the head of the patrilineage had inherited from his father, and accordingly owned. After the old man died, the main worshipping of his family's ancestors has been conducted in the home of his son, i.e. Hieu and her husband's house; although the old house still contained an altar where Ngoc worshipped. As mentioned above, unmarried women belonged to their fathers' ancestry and they were free to worship their fathers' ancestors and hope for their blessings. They could not, however, do anything to preserve the fathers' lineages, because their children would in any case belong to their own fathers' patrilineage and not to their mothers'. Therefore, when Ngoc's daughter eventually married and moved in with her husband (the most probable scenario), her mother would also have to move out, Hieu reasoned. If Ngoc remained, the house that had been in her father's family for a long time had no future. According to Hieu, the house should be kept for one of her own sons – who were the male descendants of the old man – and if none of them were able to live there as an adult, the house could be kept and maintained for worshipping purposes only. This, she had heard, had happened with a house in a neighbouring village. Nobody actually lived in the house anymore, but apparently the family gathered there for major events and death anniversaries. Female inheritors meant that the kinship continuum associated with residential land and houses was broken, and then, Hieu reasoned, it was better to leave the house without a permanent resident and use it to venerate the ancestors.

If Ngoc continued to live in her father's house, the house would be disconnected from the patrilineage. It had to be inhabited by a male heir in order to preserve its status as an integrated part of the lineage. Nonetheless, females who inherited houses were not totally unheard of. At least two other women in the village lived in their late parents' houses, with the consent of their brothers. In these cases, however, the houses were not connected to a particular patrilineage through several generations of male dwellers, because the women's fathers had not been heirs to family homes but had acquired their houses later in life. In Ngoc's case, her status as a single mother implied that she had failed in fulfilling some of the most basic female requisites, such as to get married and move in with her husband in his house, or in a house they had acquired together. Ngoc being a woman made her an unsuitable inhabitant of her current home, because Ngoc's father's house and his lineage were merged into an inseparable unity where she did not represent a lasting link.

From a regional perspective, Croll's observation that because Chinese mothers were integrated in – and not perpetually linked with – their husbands' patrilineages, they wanted to ensure their long-term security and protect their more vulnerable position by giving preference to their sons who were permanent members of their fathers' families, is illustrative. With a son as an ally, women could strengthen their positions in the family and guarantee a place for themselves in old age (Croll 1995: 95). In a similar vein, Wolf noted that Taiwanese women never experienced the innate sense of belonging that a man – regardless of how his life turned out – could feel as a permanent and unquestionable part of a family. To compensate for this insecurity and to protect their interests in a male-based family, women created strong uterine bonds with their sons. From this perspective women's relationships with their sons were shown to be crucial in order to ensure a certain level of material and social status (Wolf 1972: 160–63). As both Wolf and Croll pointed out, women were aware of their comparatively transient positions as daughters and wives, and therefore made strategic moves to protect their property interests. It can be added, just like the women in Lang Xanh, they gained their entitlement to a place to live and dwell in by their connection to a man who, as an eternal member of a patrilineage, had direct access to that place. Women's opportunities concerning access to houses and residential land as proprietors in their own right were thus inherently weaker than men's.

Men's permanence

30 years old Tien had just been called back home by his father from his job as a construction worker in the southern part of the country. He had been away for quite a few years, but now his farming parents were getting old and had begun to find it too heavy to cope with the fields, even though they had several married sons and daughters who lived in the area and gave them a hand when needed. Eventually, Tien's father

had decided that it was time for his youngest son to come back and take over the house and the farming. Still unmarried and recently returned to his home village, Tien was now on the lookout for a suitable wife. The woman who eventually married him was expected to come and live with Tien and his aging parents in their house. She was also supposed to engage in the farming and relieve Tien's mother from all the farming chores that she had been responsible for. One young woman from a neighbouring village was a candidate for some time, but one day Tien informed me that she was not going to be his wife, because she wanted to become a preschool teacher and this would not leave her enough time to take care of her share of the farming. When I left the village, there was still no wedding planned.

Tien was going to succeed his father as the resident of this house, just like his father and grandfather had done before him. The house was relatively large and of the traditional type, built of wood with beautifully carved beams supporting the roof. It was not one of the relatively square concrete buildings that had dominated for decades now. The graves of Tien's paternal grandparents were situated on the outskirts of the sizeable garden surrounding the house. Tien's status as a son made him one of the links in a row of male kin and with his moving into the house, his position was spatially manifested as well. This linkage, which connected Tien with his father, his paternal grandfather and so on, was based on the conception that they all descended from a common point of origin, an original ancestor. This idea also included an understanding of an unbroken stream of sons who were embedded in the land that contained the houses, tombs, and fields where the ancestors had lived and worked (see also Schlecker 2005: 517–18). For a peasant like Tien, whose family was not particularly wealthy, an actual or mythical point of origin was not recorded and therefore his 'source' was only known some generations back. But at this point in time, Tien's family stream had its source in this house on this land, and his future children would originate from here. Tien became a part of the deeds and movements of his ancestors that had taken place in the village and had contributed to make this a particular location for him. It was his capacity, and thus his duty, to maintain this connection between place and lineage.

Not all the families in the village were as attached to their houses as Tien's family was. Some houses were built on newly acquired ground and therefore not so tightly attached to a particular patrilineage. Usually, such houses were inhabited by men who were neither the eldest, nor the youngest, son. Therefore, these men were not the inheritors of their parents' house and did not carry any specific duties towards their patrilineal ancestors. These obligations were taken care of by one of their brothers, who preferably stayed in the family house. Such circumstances made a house less significant for the permanence and durability of the patrilineage. A location of this kind had not accumulated much of a lineage biography yet (c.f. Tilley 1994: 33). However, even families living in such houses expressed their hope that a son would establish his future family there, and show his filial piety by looking after his parents, first in their old age and later as ancestors. Each house that was dwelled in by a man and his family, constituted a part of patrilineal permanency.

Male precedence over residential land and houses

Traditionally, it has been the responsibility of parents to see to their sons' residential needs (Liljeström 1998: 88; Luong 1989: 750). By and large, this tradition still persisted in Lang Xanh. It used to be the eldest son who inherited the parental house, while his brothers, with the assistance of their parents, moved out after marriage. This usually happened as soon as it was financially feasible (see also Pham Van Bich 1999: 27). Nowadays, as Tien was an example of, it was equally common that the youngest son took over the house. If his older brothers had already moved out, it was quite customary in Lang Xanh that the youngest son stayed on with his wife in his parents' house after he had married.

Chi, an elderly widow drawing close to her 80s, remembered that she had helped three of her four sons with obtaining their houses; one of the sons had moved from the village to work at a tea company in a nearby province, but the other three lived in Lang Xanh. She was old enough to have experienced several forms of land management in the village. Now her youngest son, with his wife and two children, lived with her in the house she used to share with her husband. She planned to spend her last years with this son, and then he was going to inherit the house. The eldest son had married during the period when the farming collective controlled the land. At that time, people who needed new houses did not have to pay for residential land; they just applied at the board of the collective, and were allocated a piece of land for the new house if the board decided in favour of the application (see also Nguyen Van Suu

2004: 277). Chi had made sure that there were means available to build a cottage for her son and his family on the land he had acquired. When the second son got married, things had changed and the villagers had to pay for residential land. They could either buy land from the commune or, if there was no such land available, residential land rights could be purchased from a family who was willing to sell a plot. When the agricultural collective in the village was dissolved, some land had become available. There had, for example, been a large courtyard used for husking and drying rice, which had been turned into residential land for six households. Chi had bought the use right to one of these plots from the commune for her middle son when it was time for him to start a family. Now it contained a solid concrete building and this, she revealed, was mainly thanks to the efforts of her son and daughter-in-law.

An example from contemporary rural China shows that when a married son who was not going to inherit the house still lived with his parents, it was sometimes perceived as a sign of poverty, implying that the parents were unable to assist their sons in acquiring separate households (Wolf 1966: 71; Yan 2003: 145). The villagers of Lang Xanh held similar opinions and strove for as long as possible to avoid such living arrangements. Married brothers living together was perceived as a breeding ground for trouble and disputes. Just like in the case with Chi's middle son, it was quite common that a son and his wife themselves contributed substantially to be able to acquire a house of their own. However, the norm stipulating that parents should assist their sons in acquiring a house was largely maintained. In principle, men were supposed to be the house providers and parents commonly expressed a keen involvement in their sons' residential circumstances. If the size of the garden allowed for it, parents were often quite willing to divide their residential land and distribute it between their sons in case they should wish to build houses for themselves and their families there.

Parents with sons anticipated their family's future with a kind of certainty that was manifested in acquisitions of residential land and in house building, as the example with the old woman Chi and her three sons shows. Similar conduct could be seen in neighbouring China, where Croll noticed that it was first and foremost families with sons who built new houses in the rural areas that she studied. 'Those with sons built for a future; those without had no future' (Croll 1995: 94). She was stunned by the difference in building activities between households with or without sons, and how maxims from pre-revolutionary times lingered in such an apparent way.

When women provide residential land

Lai and Bao, an elderly couple living in Lang Xanh, had treated their daughters in a rather unusual way as far as residential land was concerned. Lai was a retired part-time nurse and Bao a retired teacher from the teachers' college in the district town, and they were comparatively affluent, with a relatively good quality house, educated children, and with connections to influential persons in the area. Lai had combined her nursing duties with some farming, while Bao had stuck strictly to his teacher's career. They had three daughters and one son, who was the youngest child; a family pattern that was not uncommon. When, after a row of daughters, a son finally was born, parents often refrained from having more children. Since the son was leading quite a successful life with his wife and newborn baby in Hanoi, his parents did not expect him to return to live in the house again. Nonetheless, it was the son who was the inheritor, with the duty to keep the house intact within the branch of his patrilineage, even if he was not going to actually live there himself. Moreover, Bao was the head of a patrilineage (*dong ho*) that originated from another province, to which he travelled on important death anniversaries, and this was going to be his son's responsibility in the future. This family had, very atypically, taken pride in providing each of their three daughters with a piece of residential land. At one of my visits to their house, Lai took me out into the road and proudly pointed out the pieces of land they had bought for their daughters. All the plots had prime locations, close to their parents' house and along one of the main roads. The two sisters who had remained in the village were both teachers and lived with their husbands in houses built on these plots. The third sister's piece of land was left unused for the time being, since she lived in Ho Chi Minh City. The reasons for the old couple's quite extensive purchases of residential land rights were probably manifold, but it could not be denied that it was a marker of prestige for their daughters as well as for themselves. The daughters had not inherited land attached to the patrilineage, as this was the privilege of their brother, but they had been provided with residential land by their parents, something that their husbands normally should have provided. This, in addition to their comparatively high educational level, set the two sisters apart from most of the other women in the village. Their position in relation to their husbands was also influenced by their possession of residential land. It indicated that their husbands had entered into marriage without bringing houses or plots. This put the men in an ambiguous position vis-à-vis their wives and their parents-in-law. The status of the two husbands was negatively affected because they had taken the commonly 'female' position of receivers. On top of this, the two men were not equal to their wives with respect to level of education. Even though they did not live under the same roof as their parents-in-law, these men were partly living in the shadow of the old couple, and as far as Lai and Bao were concerned, their daughters had married down. The daughters' husbands were rather absent from the stories that the old couple told me about their family and I seldom met them out in the village or in the fields.

Contrary to women, it was very unusual for a husband to move into the house where his wife's parents lived. Any man who did this would have his position greatly compromised, especially in relations to his in-laws but also to his own patrilineal family. A son-in-law did not have a clearly defined position in the house of his in-laws. He was not an integral part of his father-in-law's patrilineage and it was not part of his role to adjust his daily life in accordance with his parents-in-law.

Thuan, a married woman with two daughters and no sons, claimed that few men were able swallow their pride and go and live in a household where they had no obvious rights. Apparently these men were sometimes mocked and accused, mostly in a joking manner, of being like 'dogs hiding under a cupboard, afraid of their masters' blows', as Thuan expressed it (see also Pham Van Bich 1999: 28). Even though she did not have a son, she was not expecting to live with a future son-in-law in her house. Unlike daughters-in-law, who were often spoken of as acquired daughters, sons-in-law could only to a limited extent be regarded as acquired sons, according to Thuan. In her opinion, the type of arrangement where sons-in-law moved in with their wives' parents hardly ever worked out well. The parents-in-law could not fully rely on the loyalty of their son-in-law, and he, on the other hand, might find it difficult to exercise the rights and duties of a husband which would have been taken for granted if he had lived in a house that was attached to his own patrilineage. Like Pine shows in the Górale case from Poland, when a poor man from another village moved into the more wealthy house of his wife, he held an unclear position. His economic as well as his social status did not correspond to his gender. As a man who only brought his labour into the household, he took on a position that was normally occupied by the woman (Pine 1996: 451).

Defining residential land - men as house builders

As we have seen, men's position and authority as family members were closely connected to their residing on land that was incorporated in their patrilineage. Also, when it came to placing a family house on the land it was ideally handled by men. In the case of Lang Xanh, building was closely associated with a male capacity, and house building was a male duty *par excellence.*

The building of a new house was a major occasion in life for people in the village (see also Nguyen Khac Tung 1993: 38–40). It required big monetary investments and extensive organisation of all the duties involved. Women without husbands almost invariably stated house building and renovation as the greatest obstacles for a single woman to overcome. In discussions about labour division, house building was commonly at the top of men's task lists, while cultivating the fields often topped the range of female duties. A case-study of five communities in northern Vietnam, shows that along with weddings and funerals, house building was an occasion when virtually all the survey participants engaged in extensive networks of mutual assistance (Le Trong Cuc and Rambo 2001: 193). Since it was men who acted as managers of the building projects and who performed all the major tasks involved, the networks that were formed in connection with house building had men at the core in Lang Xanh.

In Lang Xanh, men's significance for the result of building and inhabiting a house was indisputable already at the planning stage, well before the actual construction work could commence. Besides providing the land, the skills, and much of the means, the direction of the house construction – according to geomantic premises – was associated with the male family head. The same applied to finding the right day to start building, i.e. picking an auspicious day in the lunar calendar to be the 'ground breaking day' (*le dong tho*). In both cases, the age of the male head of household had to be taken into consideration. In these circumstances age referred primarily to the years in the zodiac cycle, which consisted of the 12 animals. The year of birth was believed to have a strong influence on a person's fate (*so phan*)[13], and when a household had an adult male member, it was his fate that mainly determined how a house harmonised with the land on which it was built, and with its inhabitants. When the date was picked, usually in consultation with a fortune-teller, the next step was to inform the Land God and the ancestors about the new house that was to be built on the plot. Commonly they were first informed on the night before the 'ground breaking day', and then the Land God was presented with a separate ceremony on the intended building site. These ceremonies should be performed by the male head of the family, since it was his age (i.e. his zodiac sign) that influenced the timing and the outcome of the building project (see also Nguyen Khac Tung 1993: 39–41). When the construction work eventually began, women quite often assisted the male builders and also cooked meals for the men who laboured on the building site. Hence, women did engage in building, but it was always a man who possessed the overall control of material as well as of spiritual matters.

When a single woman wanted to have a house built, the procedure could be like that of Thuyen, a woman in her mid 30s with two daughters and very few assets. She managed to have a house built despite unfavourable circumstances. During my time in the village, Thuyen had her small cottage re-made into a concrete structure. She was officially still married, but her husband had not been seen or heard of in eight years. She finally had her house built despite her father-in-law's disapproval. From his point of view, she should not have undertaken such a major project without the consent of the head of the household, i.e. her absent husband. Instead, her father and three brothers had helped her to get the necessary material and build the house. Since she lacked the support of her father-in-law, it was her father who had decided the place and outline of the house and he had performed the necessary ceremonies. When she and her two daughters moved into the new house, her father consecrated the new ancestral altar on the wall and invited

13 A person's fate can be comparatively good or it can be 'heavy' (*so nang*). This resembles ideas of destiny, or fortune, and influences the outcome of a person's endeavours and aspirations. See Malarney (2002: 95) for a more elaborate consideration of the concept of fate.

the concerned deities. The patrilineage of her missing husband had thus been replaced by her father's lineage.

According to Heidegger, a building can be said to assemble and arrange a dispersed world into a significant locality, filled with intrapersonal and divine relations. 'In this way, then, do genuine buildings give form to dwelling in its presence and house this presence.' (Heidegger 1971: 155–58). For a house in Lang Xanh to hold this capacity a man had to be the possessor. In Thuyen's case, her father became the builder and the person who gathered together the surroundings with the ancestral and the divine relations into a proper house that she could inhabit. Thuyen on her own could not communicate with the ancestors and the Land God, and she needed her father to invite the ancestors to her new altar. Young concludes that: 'If building establishes a world, if building is the means by which a person emerges as a subject who dwells in that world, then not to build is a deprivation' (Young 2005: 127). Lang Xanh women were neither the builders nor the members of the household with the ability to connect a house and its inhabitants to a piece of land by consecrating a patrilineally anchored ancestor altar on the property; this was an exclusively male capacity.

A gender perspective on access to farmland

Mobile women and located fields

If we recall from chapter 2, rice fields had been allocated to the villagers around 1993, with 20 years usage rights. As the cadastral officer, Han, told me, all adults over 18 years of age – irrespective of sex – who did not have a salaried employment at that time received the same amount of rice fields;[14] they were, thus, categorised as farmers who worked the land. In this way, each farmer was attached to the fields that had been allocated to him or her for a period of 20 years,[15] starting in 1993. As long as the farmer did not move to another place or became a member of another household, the fields corresponded rather well with the needs of the villagers. Each household usually regarded their total amount of

14 Children and old people who did not have a pension from a former employment received smaller areas of rice fields, according to the cadastral officer.

15 There has not been a reallocation of the land and the revised Land Law implemented in 2014 prolonged the usage right for agricultural land from 20 to 50 years. See chapter 2, note 9.

fields as a joint asset between the household members. But when young persons got married the households' fields had to become more individualized to cover their needs. Hence, when the farmers multiplied and moved around, new situations arose.

A couple of the village men had married women from neighbouring communes. An example of this was Cam. Because Cam and his wife had married after the land allocation, they had received their fields as members of their parents' households. This meant that the fields Cam had access to were in Lang Xanh, but not his wife's fields. She still had to cultivate rice fields in her home village. The village was some five kilometres away, and she only went there for more extensive work, like harvesting or transplanting the fields. Sometimes Cam came with her to take manure to the fields or to plough them, but the daily care of the fields was conducted by her parents. Quite a few women had married and moved too far away to carry out the everyday care of their fields. As in the case of Cam's wife, it was usually the woman's relatives who were responsible for the maintenance of the fields between high seasons. But at peak farming periods, women in and around Lang Xanh, sometimes accompanied by their husbands, could be seen pushing bicycles over-loaded with tools, bundles of rice seedlings for transplantation, fertilisers, or whatever the season demanded, on their way to or from their villages of origin. If this place was too distant, like it was for Thu, who came from an uphill village some 500 kilometres to the south, all the labour was left to the woman's next-of-kin. Thu's mother always gave her large sacks of rice whenever she came back to see her family, as the fields were still formally hers, and since she had married after the fields in Lang Xanh were allocated, she did not have any land in the village. The practice of patrilocality made a woman mobile – from her natal household to her husband's upon marriage – but her fields were not movable.

As already pointed out, the tradition of patrilocal residency had continued to be widely practiced throughout the land reformations and collectivisation. Consequently, it had been much easier for men to remain close to their native land resources, despite the changes in access to farmland that the different land regimes had brought about. Before the collectivisation of land, when peasant women married and left their childhood homes, they also left the fields they had used to cultivate, in order to farm the fields of their husbands' household (White 1988: 169). The general principle was that 'out-marrying' daughters were replaced by incoming daughters-in-law, while men and land largely remained in place. This norm had basically persisted in Lang Xanh. Even though each household could now manage their farmland as they wished, the system of individual allocation of fields did not always synchronise well with either patrilocality or the formation of nuclear households – not least because when young people married and formed their own household, they were supposed to keep working the same fields that they had been allocated by the state as children. In fact, they *had* to retain their fields if they were to support themselves as farmers. In most cases, the husband's family could not support the increase in household members that daughters-in-law brought about, unless the same number of daughters moved out to work in fields belonging to their husbands' natal households, and so on. For families with more than one son, the situation was even more complex. Furthermore, in Lang Xanh all arable land had been allocated, so there was no possibility of acquiring any additional fields from the commune, or of clearing new plots for cultivation. In rare and lucky cases, some of the women had managed to exchange their fields with women from neighbouring villages.

If the state decides to reallocate the farmland in the future, women originating from other areas will probably be allocated fields in Lang Xanh. Until then, they will however, continue to be distantly attached to their remote fields and depend on their parents, or other family members, for help with the farming. It can even be argued that the land allocation indirectly strengthened patrilocality since it was virtually impossible to get access to farmland outside of one's native village. This contributed to a male dominated attachment to land over time. As we have seen, men rarely moved to their wives' home villages, and seldom did they began married life with none or only distant land assets at their disposal.

The fixed access to farmland made the situation for any newly married couple trying to make a living as farmers quite problematic, and even more so after they started to have children, since there was no more farmland available for them to support their new family members (see also Tran Thi Van Anh 1999: 110). The farmer Dat, married and the father of two small sons, summed up his family's land concerns with the following expressive remark: 'Dead people have fields and people who later became teachers have fields, but none of my two children have any fields, so in practice the allocation does not work.' A distribution of land that had been rather fair in 1993 did not, as a result of its static construction, reflect the changing constellations and needs of the households anymore. This gave room for numerous informal – both long-term and short-term – land transactions between the villagers, and to some extent with neighbouring villages too. Therefore, each household's access to farmland depended to a great extent on luck, contacts, and successful negotiations with relatives and neighbours. A very common remark, made by men as well as women, when peasants commented on their farmland situation was: 'Now, all we can do is work hard on the fields we have.' Their labour was the most controllable asset that the farmers in Lang Xanh possessed, and the individual allocation of land had not fundamentally changed this. Because the habit of patrilocal residency had remained strong, women's work was still predominantly directed towards their husbands' households and fields. From this perspective, women were even more dependent on – and obliged to make use of – their labour as a controllable asset, compared to men, who more easily remained attached to their land.

Desacralised farmland

In pre-revolutionary Vietnam, all land was perceived as more or less inhabited by spiritual beings (Jamieson 1993: 38; Kleinen 1999: 168; Malarney 2002: 41), even if the rice fields already then were less connected with spirits and the other world in comparison with residential land. However, the lines between sacred and profane land was not clear-cut and definitive, as a piece of land could be more or less potent depending on its history. As in the contemporary case of the family with the crying boy whose residential land used to hold a temple, the definition could vary over time and according to events that occurred on a piece of land. Things like accidental deaths or destruction of sacred buildings could change the character of a piece of land (see also Kwon 2008: 20–21; Malarney 2002: 43). Examples of otherworldly interventions from today's Lang Xanh concerned only residential land, whereas the rice fields were now fully placed in a secular realm.

An old couple, both in their early 80s, remembered that in their youth the farmers had performed various rituals connected to agriculture.

'Almost everything was affected by worshipping', the woman recalled. When they had worked in the fields, for example, the tools they used, like a sickle or a plough, had, at the beginning and the end of a working day, to be handled cautiously due to spiritual concerns. When the sickle, for example, was put back up on the wall after a day of harvesting, the old woman recalled that she had had to say a short prayer. This practice had not endured. The impact of the communist regime's profanation strategy (Malarney 2002: 41) was evident by the unwillingness of contemporary Lang Xanh peasants to appear to be superstitious and irrational persons, who put their trust in prayers rather than in hard work. Spiritual matters were handled in the realm of the family house and, on special occasions, at the pagoda. Some land was considered prone to harbouring ghosts or other spirits, like the land where the *dinh* had been, or places with graves. Outside those areas, none of the farmers I met with perceive the land as a dwelling for otherworldly beings, or as a site for ancestral or divine contact.

As we saw when Thuyen had her house built, the local Land God had to be contacted and he was involved in all matters regarding residential land and houses. Farmland, on the other hand, was not attached to a Land God[16]. One reason for this was given by Hao, a full-time farmer in his 50s. According to him, the absence of a Land God was due to the fact that all worshipping took place at the altar inside the house and not out in the fields. In his opinion, the Land God could only be addressed at the house-altar, and he was concerned with the families' ancestral and spiritual matters and general well-being, but not with the fields as such. Very few of the inhabitants expressed divergent opinions. However, a couple of villagers claimed that there also was a Farm God, but he could not be worshipped out in the fields or inside the homes of the farmers. According to them, this god was probably venerated at the pagoda by the most devoted old women. Bach, an elderly woman who was active at the local pagoda, was not so sure about the Farm God when I inquired about him. She told me that people followed the Buddhist manuscripts

16 Kleinen has noted that in the northern village of Lang To, people have to some extent picked up the practice of the *vao he* (beginning of summer) ritual where spirits are asked for a good harvest. He also notes that three small shrines devoted to the Spirit of the Soil (*Ong tho*) had been erected along a village road (Kleinen 1999: 169–71).

that were kept at the pagoda, and occasionally their prayers included the 'rivers and trees' and that this, according to her, really included the entire environment and, thus, also the fields. This was as far as I ever got with regard to the Farm God, and he (*ong*) did not reappear in conversations again.

As far as deities and farmland were concerned, the farmer Hao also claimed that one of the things that set the majority people (*Kinh*) apart from some of Vietnam's ethnic groups, was the fact that the *Kinh* no longer performed any worshipping in the rice fields. Ethnic people, he claimed, still left some space in the middle of their fields for the purpose of worshipping. Hao's opinion suggested some awareness of the fact that even *Kinh* peasants had, at some point, performed worshipping out in the fields[17]. In Lang Xanh, this habit had been abandoned and was now perceived by most people to be ineffective. 'The neighbours will think I am crazy' one woman said when she told me that a fortune-teller had suggested that she should light some incense and make a small offering of food and alcohol in one of her fields to enhance prosperity.

When Lang Xanh people worshipped in their homes or at the pagoda, a standard request was for the family to have 'good business' (*lam an tot*) (see also Malarney 2002: 100). I persistently and extensively tried to find out if the farmers in the village ever made a whish for a good harvest when they worshipped their ancestors or the deities. But despite my efforts, I could not find any evidence that indicated this. The 'business' that people included in their wishes referred to everything that the household members engaged in, in order to support themselves. As a result of my inquires about this, I was frequently told that it was not the worshipping of spiritual beings of any kind that gave farmers a good crop; it was a combination of good seeds, luck with the weather, and spending enough time working on the fields, that characterised successful farming. Indeed, I often heard farmers, especially women, commenting on the negligence of some farmer, most often also a woman, to put in the amount of labour that was required to improve their yield. Hard

17 Malarney describes how the fields of pre-revolutionary Thinh Liet (a northern Vietnamese village) had altars at the edges where the peasants, through rituals, asked the spirits for permission to work the land and wished for a good harvest. He also explains that the entire landscape was divided into sacred and profane areas, even though the distinctions were sometimes fluid (Malarney 2002: 41–44).

work had entirely replaced prayers and offerings as a way to achieve good agricultural production.

Impacts of the state on land control

Already during the colonial period, the elite from which the leaders of the Communist Party were being recruited called for reformation of ritual practices. In 1946, soon after the proclamation of the revolutionary government (August 1945), campaigns to downscale and reject rituals and ceremonies were launched (Malarney 1996: 540–41). The Communist Party harboured strong scepticism towards any ritual or spiritual practice that might enhance the power of a hierarchal society based on traditional patriarchy. For ideological reasons, and as a reaction to a strictly stratified, Confucian-influenced, society, the government saw 'superstition', which meant all practices aimed at involvement in the otherworldly (e.g. spirits, fortune-telling, and auspicious dates based on the lunar calendar), as an obstacle to progress (ibid.: 540–42). Rituals, such as funerals, weddings and ancestors' death anniversaries all included a certain amount of dealings with the spiritual world, and as such they became cultural obstacles that obstructed the regime's intentions; namely to build a secularised state which repudiated patrilineally informed loyalties and 'irrational' behaviour that slowed down production, and instead propagated for a society based on solidarity among proletarians and a rationalistic worldview based on scientific knowledge and technology (e.g. Goodkind 1996; Luong 1993; Malarney 2002).

Seen from a land perspective, the purpose of the communist regime's land management has been to break the entity that was formed around patrilineages, spiritual beings (including ancestors), and land, and which contributed to connect patrilineal families with property and particular places. Traditionally, the tombs, graves, ancestor halls, and inherited family homes had attached patrilineal families firmly to the villages and the surrounding rice fields. With concrete changes in land management as a tool, the revolutionary state's intention was to abolish people's trust in the supernatural, in order to alter the peasants' relations to the land they cultivated and inhabited (Malarney 2002: 50–51). If the links between the peasants, their ancestors, and the spiritual world could be weakened, the relation between kinship ties and land would be destabilised and new norms could more easily be implemented, the state believed. Despite their efforts, however, the regime never entirely managed to separate the worldly from the otherworldly, as the examples from Lang Xanh people's dealings with their ancestors and residential land have shown. The cult of the ancestors was still present, regardless of reformist winds that had, with varying intensity, encompassed ritual life since the revolution (Goodkind 1996: 729;Kleinen 1999: 161; Malarney 2002: 92; Pham Van Bich 1999: 224). The secular was entangled with the sacred, and since these worlds were to some extent perceived by the people as interdependent, an integral part of the rituals and of ancestor worshipping was to keep the relation between the two worlds harmonious. If families refrained from caring for their ancestors and the deities, people would suffer a loss of prosperity, which was assumed to be the reward of their ritual efforts (Goodkind 1996: 721; Malarney 2002: 80, 87, 91).

To commemorate the forefathers was, and still is, a deeply rooted Vietnamese tradition, and this left the communist regime no other choice than to condone the continued practice of honouring and remembering ancestors. After all, the communist cadres also had ancestors and families with the same tradition. What separated the state's view from people's understanding of ancestors, was the resistance of the former against any dealings with the otherworldly. The regime had wanted the rituals involving ancestors to be of a purely commemorative and honouring kind. This was in direct contravention of the practices and convictions involved in ancestor worship, since the ancestors dwelled in the other world (Malarney 1996: 542, 544, 2002: 92).

The detachment of peasants from farmland

Before the land reform (1953–56), fields – predominantly rice fields – used to be connected to the sacred sphere in at least two ways; they were the loci of spiritual beings, and they provided means for ancestral and spiritual purposes. The first point was, for example, manifested in peasant ritual practices, where the deities or spirits were addressed in order to wish for a good harvest or to get permission to work the land (Kleinen 1999: 169; Luong 1992: 56; Malarney 2002: 42). The second point was evident in the fact that rice fields owned by family clans, or by village associations devoted to worshipping village deities, were used to sustain worshipping practices by making sure that there were resources available for their rituals and worshipping (Nguyen Tu Chi 1993: 71;

Phan Dai Doan and Nguyen Quang Ngoc 2002: 38–39; To Lan 1993: 177–80). It had also been possible for heirless people to donate land to the local temple, or leave it in the hands of an 'adopted' family to cultivate. This was done to ensure that worshipping and death anniversaries were conducted even without male heirs, and that the expenses for these events were covered by income from the land (Malarney 2002: 44; To Lan 1993: 178–79). As a direct result of the drastic changes in land control brought on by the land reformation, and as the examples from Lang Xanh showed, none of these practices had continued. After the land reform and collectivisation, none of the farmland designed for sacred purposes had been left intact. All such land was confiscated and redistributed as regular fields. Consequently, the fields' symbolical and substantial value for the continuation of patrilineages' ancestral cults and communal worshipping vanished.

According to Tilley's idea about the relationship between people and land, the land with which people form the most durable relations, is the kind of land that enables people to feel an unbroken connection to those 'powers and figures' that established their relation with a particular location. These relations, in turn, are maintained through ritualised procedures (Tilly 1994: 11, 14). It can be argued that residential land in Lang Xanh held such qualities and these had endured through reformations and collectivisation. The peasants' relations to farmland, on the other hand, had developed in a different direction. One of the reasons for this was the Party's wish to create an agricultural sphere that was characterised by effectiveness and high yields, produced by specialised labour brigades. The ideology propagated by the party at one time stipulated that:

> When the masses still believe in the heavens, spirits, and fate, they will be powerless before natural changes and the difficulties that the old system has left behind. People will not be able to completely be their own masters, the masters of society, nor the masters of the world around them (Ninh Binh, Cultural Services 1968: 6, cited in Malarney 2002: 80).

Therefore, the regime attempted to redefine the meaning of land, particularly farmland. Farmland was a resource that should be worked by the peasants in a fashion similar to any other resource utilised by labourers in national production (Malarney 2002: 49). The most important step in this direction was abolishing private ownership of land. For that reason, the state declared itself as the sole owner of all land when collective farming was implemented in the late 1950s. This new land management excluded a continuous relation between the farmers and any particular fields, and the crops they produced were appropriated and distributed by the state.

The attempt to dissolve emotional bonds between farmers and specific plots of land has been a common feature in places where collective farming has been implemented. In Bulgaria, Kaneff notes that this was a central strategy in order to redirect the farmers from landholders into workers (Kaneff 2002: 36). The emphasis on labour as the ideological base for citizenship, and not on attachment to particular pieces of land was also a dominating trait in the Soviet Union, as Perrotta writes in an account from post-Soviet Ukraine (Perotta 2000: 168).

In Vietnam, the large-scale and industrialised mode of production, where the peasants' labour efforts did not appear to them as having any direct or predictable correlation with the actual outcome of the farming collective, or with the revenues they got for the work, resulted in unmotivated labourers. Chu Van Lam captures the atmosphere aptly when he writes that 'labourers who once cared for the land as they would their own bodies began to neglect it as something foreign to them' (Chu Van Lam 1993: 156)

Undoubtedly, the collectivisation of the peasants' land and labour separated the farmers from particular fields to such an extent that it was impossible to invest them with spiritual power, or any kind of symbolic significance for the preservation of patrilineal families, and this was what the government had intended.

Degendered farmland

The Communist Party's idea that farmland should be incorporated into the sphere of resources for national production implied that these places should be occupied by labouring citizens, i.e. by men and women alike. In order to achieve national production goals, it was regarded as necessary to include women as well as men in the labour force. At the same time it was assumed that women's equal participation in labour led their way to emancipation, as they would be able to compete both socially and economically with men and, in this way, become their equals (see e.g. Mai Thi Tu and Le Thi Nham Tuyet 1978). It can be argued that the intention of farm collectivisation was to turn the fields into a public space, where women were supposed to transform themselves into equal workers, despite existing norms of patrilocality and women as family caregivers. It was assumed that women, in the shape of labourers, should be let in on an equal footing with men, on what was suddenly intended to be gender-neutral ground.[18] But the fields had never been gender-neutral. On the contrary, the gendered division of labour and control of land had, since far back in time, been quite distinct (White 1982: 45).

As an example, before the land reform, privately owned fields had been passed from father to son, where the eldest son commonly got an extra share of fields, called 'incense and fire fields' (*ruong huong hoa*), to compensate for the costs involved in his future worshipping obligations towards his parents and other ancestors (Luong 1992: 74; Nguyen Tu Chi 1993: 58; Pham Van Bich 1999: 26; To Lan 1993: 183; Jamieson 1993: 24). According to custom, the eldest son also inherited the land that held the graves of the family ancestors. This piece of land was devoted to ancestor worship and was protected by customs so that it could be kept intact by the eldest son, who, therefore, was tied firmly to the land in this place (Luong 1992: 74; Wiegersma 1988: 54). In pre-revolutionary Vietnam, because of the male-biased practices of inheritance, women normally did not have access to farmland in their own right (White 1988: 168). Yet, there are exceptions to be found. In the south-central province of Binh Dinh, a land register from 1815 showed that in one village, 30 per cent of the land owners were women; unfortunately the author does not elaborate this fact any further (Phan Huy Le 1993: 200). The legal codes of the Le dynasty (1427–1783) are sometimes pointed out as relatively, even uniquely, protective towards women's rights to property[19]. However, Nhung Tuyet Tran has examined the Le

18 In *Changing Identities of Chinese Women: Rhetoric, Experience and Self-Perception in Twentieth-Century China*, (1995), Elisabeth Croll observes the gap between women's actual experiences of their possibilities to become equal members of society, and the rights and positions that rhetoric attributed to them. Very few of the traditional male dominated spheres were in reality opened to women in socialist China.

19 The Le code has been used as evidence of Vietnamese women's unique position in society as compared to for instance China, and also as evidence of early elements of gender equality and, consequently, a breeding ground for socialism and solidarity (Nhung Tuyet Tran 2008: 43–46).

code and found that only in exceptional cases could women – normally in lack of brothers or sons – who had access to property gain the right to control some of it until their death. This normally happened through donating property to a temple or pagoda in the village, in return for a promise that the institution would perform ancestral rites for her for eternity. After a woman's death, the property went back to the patrilineal family of her father or husband (Nhung Tuyet Tran 2008: 59). In her analysis of the Le code, Nhung Tuyet Tran found that in effect it clearly supported patrilieally based male succession and inheritance (ibid.: 66). In the northern part of Vietnam, the collectivisation in the late1950s ended all traditional procedures for access to farmland.

The situation in contemporary Lang Xanh indicated that by allocating farmland in accordance with the stipulations of the Land Law (1993), the state had, in a very tangible fashion, taken on the role of provider and distributor of farmland; a procedure that formerly used to be handled by the male heads of the households. In this sense, men and women now became equals; gender did not matter when the state allocated fields to the households, and female and male farmers now acquired the same amount of farmland. And, as has been pointed out earlier in this chapter, this differed greatly from residential land, where the changes in practices of access had been relatively few. As for farmland, however, men were now put in the same receiving position as women traditionally had been.

Persisting ties between patrilineallity and residential land

As we have seen, the focus of the first land reform (1953–56) was predominantly aimed at farmland, where the confiscations and redistributions were extensive. To my knowledge, the families in Lang Xanh were spared from having their houses confiscated during this reform. In other villages, however, some of the people classified as 'landlords'[20] did not only lose their fields but also their houses, or parts of them. In the process, possessions such as ancestral altars and other family heirlooms were sometimes lost. This too was part of the intended assault on the power of the former patriarchal elite, which was manifest for instance in their comparatively large and elaborate houses and altars. In the following 'Rectification of Errors' campaign, some of the wrongly classified

20 See Hy Van Luong's account of the classification of households and the process of land reformation in Son-Duong 1954, not very far from where Lang Xanh is situated, for a thorough insight into the effects of the reform (2010: 166–78).

families got their houses back (Malarney 2002: 33; Moise 1983: 213, 251). Later, when collective farming was implemented in the northern part of Vietnam (1958–61), also residential land was formally managed and controlled by the state. This, of course, involved Lang Xanh as well. When this happened, the villagers by and large continued to live in the houses they had occupied before the collectivisation, so, in practice, the changes were relatively few (see also White 1988: 170). In some cases, though, the collective board had allocated parts of people's gardens to other families who wished to build a house of their own. Some villagers remembered that the gardens of certain houses used to be much bigger because they once had included what was now somebody else's residential land. After the collectivisation ended, some of this land had been restored to their former owners.

On the whole, it can be concluded that houses and residential land in Lang Xanh had by no means been exposed to fragmentation and redistribution to the same degree as had happened with farmland. Consequently, it had been possible for the families in the village to uphold not only inheritance practices, but also basic residential principals; i.e. patrilocality and the preference for male heirs of houses and residential land. As an example, when a couple got married in Lang Xanh during the collective period, the woman just transferred to her husband's unit, if they were not already in the same one. In this manner, the wives continued to move to their husband's household, and the sons adhered to their duty to live with their aging parents and take on the responsibility of ancestor worship. With the house as a base, patrilineally oriented families could largely continue their residential, marital and some of their ritual habits.

Kerkvliet (2005) claims that the Vietnamese peasants showed their discontent with the agricultural collectives not so much by open resistance as by directing their efforts towards their social networks and the small land plots outside the collectives[21]. Most likely, this priority added to the maintenance of the patrilineal household during the collective

21 The small plots that the households could keep during collective farming were commonly referred to as 'five per cent land' because it constituted about five per cent of the arable land belonging to a farming collective. There were several reasons why this land existed, one of them being that the collectives were unable to provide everything that was needed to sustain the members. The party also assumed that it would keep the peasants in a more benevolent mood if they could,

period. The fact that peasants largely continued to live in partilineally oriented houses and households, and could keep small plots of land that they did not formally own, yet still controlled to a great extent, has most likely had a tremendous impact on the endurance of the rural household against state ideology. Even while the government's aversion against rituals and worship was at its peak, these practices could prevail in and around the house.

The importance of the residential house as an outpost of resistance against socialist movements have been noticed by Frances Pine among the Górale peasants in Poland. When the state tried to absorb the households, in their capacity of economic as well as of ritual units, the Górale resolutely adhered to their domestic rituals and could, in this way, preserve the symbolic importance that the houses provided for the family clans (Pine 1996: 456). In her study of the meaning of the house in the tensions that arose between the Polish state and the Górale moral economy – that was based on ritual and kinship – she found that a main characteristic of the house was its ability to endure against ideological as well as more actual encroachment from a dominating state apparatus. Without being static units, the peasants' houses constituted main loci for passive resistance during troublesome times (ibid.: 445).

In Lang Xanh, comparable house-based endurance against state intervention could be noticed. Present day land legislation did not force, nor actually encourage, the villagers to change their inheritance practices as far as houses and residential land are concerned. The Land Law (1993) does not distinguish between men's and women's rights or responsibilities; it addresses the land-users as 'domestic households and individuals'.[22] This formulation implies gender equality as far as land access of any kind is concerned (Werner 2002: 36), but does not explicitly call for any changes in access and inheritance practices concerning residential land. Contrary to the allocation of farmland, as far as access to residential land is concerned, factors like gender, kinship, and marriage practices are still decisive factors. Moreover, in contrast to farmland, which has been allocated for a period of 20 years (now prolonged to 50 years), residential land is assigned to the users without a fixed time limit.

at least initially, continue to farm a small share of the land according to 'custom' (Kerkvliet 2005: 76).

22 Vietnam Land Law No. 45/2013/QH13, Chapter 1, Article 5, Section 2.

This the villagers interpreted as a use right that, in practice, would last more or less eternally.

Gender and place – a discussion

Gendered access to residential land and houses

In Lang Xanh, as we have seen, the connection between residential land, patrilineages, and patrilocality, had been kept intact and this continued to put men and women in different positions in relation to both kinship and place. Hence, a place-based core, where the patrilineal family, the altar and the house on its plot of land formed a unity, was significant for the establishment and well-being of the family homes in Lang Xanh. In order to gather together the human world in the sense referred to by Heidegger (1971) – i.e. to turn the unity of a house and a piece of land into a meaningful location by connecting the dwellers with each other as well as with the otherworldly – a house not only required the land that contained it but also a household headed by a man.

It is evident that part of the male authority in a Lang Xanh household rested on these perceptions and on the idea that a man was the unquestionable proprietor who had provided his wife and children with a house and a piece of land. When men did not live on land that could be attached to their patrilineage, like self-obtained land, or land received from their parents, the equilibrium between male and female positions was destabilised. This had happened in the case with Lai and Bao, who had bought residential land to their daughters, which, in turn, had put the daughters' husbands in socially precarious positions. Men were not supposed to bring only their labour into a marriage; this position was preferably occupied by women. As the requirements for Tien's future wife suggested, labour was one of the most important assets that a women could contribute to the household.

If we follow Ingold's line of thought, that people and places are distinguished through their relations and positions *vis-à-vis* each other and the people who have previously inhabited a place, it can be argued that Tien came to be who he was through his interactions with his predecessors in this specific location. When he inherited the residential land, with the house belonging to it, and took up his living there, Tien positioned himself within a network of patrilineal relations that constituted both him and this place (Ingold 2000: 133, 149). The features of the place contributed to make Tien the man he was, at the same time as they composed a habitat for Tien to form in his particular way. It also follows that the woman who becomes his wife will be positioned in relation to the piece of land and the house on the basis of very different premises. Tien's connection to the land and the house was derived through his indisputable position in the patrilineage. His future wife, on the other hand, will live there because she is incorporated into a patrilineage that has resided there for several generations. Her presence will be of a more temporary character, and will be endorsed by her relation to Tien – and the labour she provides. Even unmarried and, as yet, without children, Tien was nevertheless upholding his lineage and, unlike the single mother Ngoc who could do nothing to preserve the connection between the patrilineal family and the house she occupied, he embodied the ability to maintain the house and its connection to this place. If the house had a male head, then that house was connected with an eternal flow of related men (c.f. Schlecker 2005: 517).

Women did not constitute permanent components in either the perpetuation, or the uniting, of worlds and beings. A daughter, who was supposed to leave her natal family to go and live with her husband, could not provide her parents, or her father's lineage, with the same level of comfort and security as a son could (c.f. Gammeltoft 1999: 71). Where and how their sons lived was therefore of greater concern to the parents, since their old age and the future of the patrilineage depended on the sons. To provide a son with a house was a way to hold sway over his loyalties, and to create a certain kind of stability under inevitably changing circumstances (see also Carsten 2004: 34–35). From this perspective, house and land was not only things to occupy, they were, as Carsten argues, factors involved in the making and embodying of relational hierarchies (Carsten 2004: 35, 37). Male and female dwellers found their respective positions in relation to each other, and the location they inhabited, based on differing premises.

Directing women's practices towards the patrilineage

As this chapter has shown, most adult women's duties included the everyday maintenance of their households' relations with deities and ancestors in order to ensure the comfort and safety of the family members. This role was further enabled by women's more peripheral positions in local as well as national political interests – where otherworldly matters were frowned upon – which offered them relative independence to act within the comparably precarious spiritual sphere (Malarney 2002: 102). Thus, even without official support and despite a rather widespread male reluctance to practice more than a minimum of worshipping, contact with the spiritual world could be maintained through female practices. Hence, women participated in forming and preserving the relations between the patrilineage and the otherworldly within the place of the house. An example of this was the case of Loan and the boy's mother, when they tried to calm the crying boy by creating harmony between the ancestors of the male head of household and the place where they lived. As mentioned, every act of worship required male involvement, for instance, a woman stated her husband or father's name before she could proceed to revere the ancestors. Through practices like these, the male-oriented kinship system directed women's house-based worshipping towards the maintenance and preservation of the patrilineal family (c.f. Li 1998: 677;Werner 2002: 38).

The unmistakable demarcation between male and female dwellers in a place was illustrated by the 'outside' position of Sinh's (who was sure that her dead mother's soul would be warded off the ground by her husband's ancestors) natal lineage and her husband's 'inside' patrilineage. This made Sinh's position as a wife, mother, and dweller in this house apparent, not only in her relation with this world, but also in relations to the other world. The change of direction of her loyalties at marriage, from one patrilineage to another, was manifested every time she worshipped. In this way, the practice of ancestor worship continuously informed Sinh's relation to the house she lived in as well as to her husband and son's patrilineage. The ancestors, the altar, and the house with its courtyard, formed a patrilineal unity centred around her husband. Through her marriage, Sinh was now a part of this, but not her mother. Sinh and her husband were oriented toward this place from different directions. These differences were manifested in their relations to the entire span of ancestors, spirits, generations, and place, all drawn together in the house and the site that contained it.

Without adopting a gender perspective, Heidegger claims that dwelling consists of building and preserving (Heidegger 1971: 145). Young, wanting to apply a feminist view on house and home, takes Heidegger's ideas on dwelling as one of her vantage points. She notes that at an early stage in his reasoning he deserts preserving, or cultivating, in favour of building, and this gives men privilege over women as dwellers. As Young notes: 'On the whole, women do not build.' (Young 2005: 126). She continues by arguing that in a male-biased gender system, men perform the building, while women become the nurturers of the male builders as well as of the buildings they erect (ibid.: 130). Young suggests that to be a builder, or to be someone who occupies a building, are two different ways of dwelling. Since the act of building is valued higher than the preservation of houses, this contributes to a patriarchal system where women's opportunities to be active subjects fall below those of men's (ibid.: 126–27, 130). From the Lang Xanh examples, we can see that women's labour and engagement in ancestor worship not only helped to preserve the houses and the male builders, but also sustained the patrilineally based family and patrilocallity. Men in Lang Xanh did not only build houses, they built unbroken family ties that comprised the living as well as the dead and the unborn. Men dwelled in their capacity as building blocks of continuity and therefore had direct access to their dwelling place. Women dwelled by way of men and the ancestors of these men.

Gender and access to farmland

Farmland was no longer regarded by the villagers in Lang Xanh as a spiritually powerful element; it was defined as comparatively public space, devoted to profane activities sanctioned by the authorities, and it did not inflict on peoples' lives on a spiritual level. The success of the regime's efforts to secularise the citizens and induce a 'rational' and detached attitude towards land had, however, not been entirely effective when it came to residential land.

Both Tilley and Abramson argue that land that has been propertied loses much of its meaning in an ontological sense, because it becomes separated from the history that has been accumulated over time, when it is transferred from one holder to another as any other type of commodity. In its shift into a marketable commodity, the main purpose of the land changes from significance and durability, to a useful means for production (Abramson 2000: 8–18; Tilley 1994: 21). In many ways this resembles what had happened with farmland in Lang Xanh. The greatest rupture between farmers and fields occurred when all farmland was incorporated into collective farms. Now, when collective faming had come to an end, the farmers had been allocated fields on a temporary basis, but their relation to this land was mediated through the state's ownership and final control. Therefore, quite contrary to residential land, which held a central position in tying men and women to particular places through marriage and kinship, formal distribution and control of farmland was now largely disconnected from kin structures and gender relations. Male and female peasants had access to arable land as a production means, and it was their status as farm labourers that legitimised their access, not their social relations.

Hence, formal access to use rights of farmland was now gender-equal, and so was the farmers' lack of ultimate control of the fields. From this perspective, men had been downgraded and, for the first time, men and women experienced similar limitations in power. Women had, to some degree, advanced their positions in their capacity as holders of land use rights. However, since the state held the ultimate control over decisions and distribution of farmland, women had not gained as much in position as men had lost. Men were deprived of the right to own and control farmland, and so were women. In other words, women were not let into this previously male-dominated domain, but men were excluded from what used to be their privilege.

Because of the detachment of peasants from the fields, farmland had become separated and disengaged from private, male inheritance and control to such an extent that a gender-equal distribution was possible in the latest land allocation. We could say that the fields had, to a great extent, been 'degendered' in the process of forming a totally state controlled management of farmland. This, however, had not necessarily resulted in the ideal gender-neutral farmers that the regime once hoped for. This will be exemplified in subsequent chapters, where men and women, in the context of farm work, engaged with the fields in ways where notions of gender-specific traits continued to be practiced and negotiated.

CHAPTER 4

'Big work' and 'small work' – gender and labour division

As a newcomer to Lang Xanh, one of the first things I became aware of was how influential gender was on the division of many of the tasks in farming. Gradually it became clear to me that the rice fields surrounding the village were far from gender-neutral places. This was especially true when it came to wet rice cultivation. The farmers often referred to 'male' and 'female' duties in very commonsensical ways. A gendered labour division tended to appear as natural and inevitable to women as well as men. When male and female bodies practiced their specific skills in their particular locations this could, according to the villagers, be performed in few other ways under the prevailing circumstance. Much in line with the ideas of for instance Bourdieu and Moore, the farmers in Lang Xanh saw their divided practices as innate to their situation because gender differences actually materialised themselves in tangible items and places that made them appear as natural (Bourdieu 1977; Moore 1994: 72, 1996).

As this chapter shall explore, the actual chores that the farmers performed did more than inform about what tasks women and men engaged in. Gendered division of labour also conveyed ideas about how 'male' and 'female' characteristics ideally related to and complemented each other, and gave an insight into processes in which men and women incorporated current understandings of femininity and masculinity available at this specific time and place. The implications of a gendered division of labour expanded to encompass ideas about the appropriate – and some of the assumingly innate – characteristics of femininity and masculinity, and made the close connections between perceptions of labour and of discourses of 'proper' femininity and masculinity clearer.

Vietnamese men and women have a long tradition of labouring the land. As pointed out in chapter 1, labour is an activity that has received much attention from national doctrines as the duty *par excellence* whereby people – especially in the rural area – were meant to show good morals, loyalty and industriousness towards not only fellow family members and villagers, but also in relation to the state (see also Bélanger and Barbieri 2009: 19, 33; Pettus 2003). This chapter will discuss the significance of cultivating morality, especially for women, while engaging in farming and the division of farm labour.

Farm work routines

In Lang Xanh, the largest share of farm labour was performed by women as already indicated. Men also farmed, but for many of the male inhabitants, the main focus was on other, more lucrative, types of work. They were, as mentioned before, engaged in building and construction projects, worked as teachers, cadres, and factory workers, or tried to find other salaried occupations, something that often took them out of the village for long periods of time. Women were at the core of farming and in practice, they not only performed but also organised most of the activities. Within the sphere of women's labour, farming was the most significant female activity and complemented, as well as compensated for men's engagements in other work. Even though the status of farming, from the perspective of the state, had declined considerably, in the local community over the last decades, women's competent farm work could be a source of respectability and appreciation. The appearance of the female farming body might diverge from the ideal of smooth and pale feminine elegancy, or from the educated wage earner, but the ability of the Lang Xanh women to handle farm work was central for them – and the opportunities to present their competences were abundant.

In high seasons, like the transplanting of the winter rice crop in January, and in June, when it was time to transplant the summer crop, women worked from early morning to late evening, only having a break for lunch in the middle of the day. 'Women only take lunch naps during summer' one of the female farmers commented. She was referring to the hot weeks between late June and late August, or early September, when the summer rice crop was ripening in the fields. Then women could enjoy the luxury of some rest after lunch, when the heat suggested it was better to stay inside if it was not absolutely necessary to go out. During the periods of transplanting or harvesting rice, women's workdays often

Figure 5: Summer transplanting

started around six a.m. when groups of women gathered in the fields to start their work. The field holder was usually the first to arrive and the last to leave, eager to get the work done as swiftly as possible.

Despite the long and strenuous days, many women asserted that much fun was involved in working side by side with other women who were relatives, friends, and neighbours. The working groups usually consisted of two to 12 women, depending on the size of the field and how pressed for time the field holder was.

When the group of women took a rest from the backbreaking work of either transplanting rice seedlings, or cutting the ripe rice with sickles, they gathered and chatted, laughed, and joked together. These little breaks were sometimes taken in the shade of a large tree, or just sitting on the banks between the fields for a moment. Sometimes travelling ice cream vendors patrolled the area during the hottest working periods. They were commonly men who pushed bicycles fitted with boxes filled with little plastic bags of homemade ice cream that they offered to the labourers in the fields. Usually the field holder treated her female co-workers to this refreshment. When it was time for lunch, each woman normally went home to her own household to eat. If she was lucky, her husband, her mother-in-law, or a teenager free from school, might have prepared lunch. Otherwise, this duty also fell to the farming woman. After some more hours in the fields, the women returned to their homes in time for taking a bath and eating, probably also preparing dinner. Even at these most busy times, there was laundry to do, children and elderly to care for and pigs and chickens to feed. All this was basically regarded as women's duties, even if other household members to some extent contributed to performing these chores as well.

When girls were between 14 and 16 years old, they started to learn how to transplant and harvest rice. Depending on their studies, if they continued schooling and went to high school, which many did, their participation in farming was usually not very strict. During the first years, girls participated mainly for fun in transplanting. If it was possible, most parents prioritised their children's education and they also held the idea that children should be allowed to be children before they started to work. 'Some have to start when they are nine years old, other women never need to learn how to transplant', as one woman described it. While I was in Lang Xanh, young boys and girls occasionally paid visits to the fields or played near to them, but they did not work. Most of the women who worked in the fields were between 18 and 60 years old, with newly wedded young women and women who had not yet started to have grandchildren constituting the core group. The organisation and recruitment of fellow farm labourers will be further dealt with in the next chapter, but it can be said that younger daughters-in-law provided a good share of the farm labour at the same time as their mothers-in-law gradually withdrew from farming. 'As soon as I married my husband, I knew I was going to be a farmer' one woman remarked. We had chatted about the hopes and dreams of young girls and she had harboured some ideas of becoming a preschool teacher; or at least, to do something different from the hard work of farming. But she had married relatively young, only 19 years old, and her husband's family had fields to farm, so her destiny was sealed. Now she spent many of her days engaged in some kind of farm related chore, accepting her 'fate', as she said, despite her previous desire not to be a farmer.

When the farming season was in a lull, the fields still needed to be tended to. Women made sure that the water level in the rice fields was right. They applied fertilisers or manure by hand from trays woven from rattan, spread pesticides by carrying containers on their backs and spraying the fields, and weeded when necessary. These tasks were not strictly performed by women, it certainly happened that men were seen out in the fields doing some of these things, but mostly it was done by women. In low season, women went alone, or sometimes in couples, to pay their fields a visit and make sure that everything looked fine. This was a duty, but also a bit of leisure. A woman could often be seen strolling to and from the fields, quite often accompanied by another woman. When they went and 'visited their fields' (*di tham ruong*), they at the same time got an opportunity to leisurely move around in the area. Men also went to visit the fields, but they did not need to relate the purpose of their movements to a particular duty or place to the same extent as women did.

Twice a year, in preparation for transplanting, the rice seedlings were sown on smaller, dry field areas, normally some square metres in size. When the seedlings were ready to be tied into bundles for transplanting, they were about 20 centimetres high with an almost radiant light green colour. This was not a strictly gendered task either, both men and women could be seen tending to the seedlings and quite often, men helped to transport the bundles to the prepared fields, either by shoulder poles, or carts pulled by themselves or with the help of a cow or a buffalo.

The preparation of the fields, especially the rice fields, was the farm chore most closely associated with men. This was, to some extent, the male equivalent to the women's work of transplanting and harvesting rice. As soon as the rice was harvested, in May and August/September, the fields were immediately prepared for a new rice crop if it was May, and a crop of corn if it was autumn. It happened that some fields were hoed by hand instead of ploughed by buffalo or cow. This was either done on smaller fields or if the household had not been able to get hold of a ploughman and a draught-animal, due to lack of money or late recruitment. Most of the households did not have a cow or a buffalo to work with and they normally employed one of the around ten men and one woman, who had access to an animal and possessed the skill to plough and harrow. Hoeing could be done by women as well as men, but ploughing needed to be performed with the help of buffalos or cows. Ploughing and harrowing were valued skills and the men who engaged in this took pride in their work, as did the woman who performed the same tasks. Since many people needed their services at more or less the same time, these men and their animals could be seen out in the fields from early morning and into the evening in order to meet the demands for their services. The age span of the ploughmen corresponded rather well with the transplanting women, even if men who ploughed other people's fields for payment tended to be at least 25 years old, and they could keep on until they felt that their bodies were too old to continue.

Sometimes during the transplant period, and usually at harvest time, men were engaged as carriers. Even if women could do this too, it was regarded as a male duty. At least one man, but often two, three or more, depending on the size of the fields, had to be recruited to carry, either bundles of seedling to the fields, but most commonly, harvested rice from the fields. While women cut the rice and made them into sheaves, men collected the sheaves and took them to the house of the field holder. Carrying manure to the fields, which was very heavy if this was done by shoulder pole, often fell to the men's lot.

If we picture in our mind's eye a view similar to the one Bruegel painted of the wheat harvesters, but this time in Lang Xanh in the month of May, we might see a wet rice field where a man and his buffalo are ploughing the heavy dark brown-greyish soil. There would be no other people in the field. He is working alone, striding ahead behind the animal with his back straight and his hands on the handles of the plough, going back and forth, making furrows as he prepares the ground for transplanting. Harrowing is carried out in a similar fashion; a man with a harrow behind a draught animal, methodically evening out the furrows as the final preparation of the field before transplanting. On the following hot summer day, the field would probably be occupied by at least six women if it were a reasonably large one. The women gather at one side of the field and start working side by side, normally within a meter or so of each other. They are equipped with a bundle of rice seedlings in one hand and with the other hand they form little bouquets of seedlings and plant them evenly in neat rows, which gives the newly transplanted rice field a rather symmetrical, lined pattern. This type of work keeps the women's bodies bent towards the water-filled field for several hours at a time. They only straighten themselves up occasionally to wipe the sweat from their foreheads, to exchange a quick word with a passer-by, or to get more seedlings to transplant. As each woman gets more and more

Figure 6: Harvesting the first rice crop of the year

experienced, she keeps pace with the more skilled women and gradually they will be fairly synchronised and coordinated in their movements. They follow each other as they slowly move backwards with their hands in the watery mud, their heads facing the ground and their backs curved. Lan, my landlady, told me that when she first started to learn transplanting, she had planted five bundles the first day. An experienced woman could transplant up to 100 bundles in a day.

If it instead had been at harvest time, as in the Bruegel painting, the scene could again show a group of women with their backs bent and their hands busy, this time with a sickle in one hand and cut rice stalks in the other. This time, the women move forward, working their way side by side until the field was empty. Most likely, there are a couple of men moving back and forth to fetch the sheaves and carry them back, either on their shoulder or by shoulder poles, to a waiting cart. They move one by one, staggering from the weight of the sheaves on their way to the cart and tread more lightly on their return.

The work that men and women did in and around the fields formed their bodies. The muscles and joints on female and male bodies were toned in accordance with their movements, and their skin darkened where the sun could reach and thickened where the materials and tools rubbed it. Both women and men in Lang Xanh were consciously observant of the shape and size of their bodies and I engaged in many conversations about the differences between bodies in, for example, the city compared to the countryside, farming bodies compared to less hard working bodies and, not least, my body compared to their bodies. Women and men often described themselves as small and thin with dark and rough skin – typical farmers' bodies according to them. When they were outside of their familiar milieu, some of them told me that they felt that their bodies showed that they were farmers because of the visible signs of the labour they did and the conditions of the places where they did their work. Inevitably, a woman who married a man with access to farmland – like the woman who had hoped to be a preschool teacher – would acquire a body that was oriented towards the fields and the farm work that women did here. And so would the men, from their specific work.

'Big' work and 'small' work – a dominant discourse

When I began to look at gender aspects of farm labour division in Lang Xanh, I had already countless times heard that men do 'big' work (*viec lon*) and women do 'small' work (*viec nho*). As far as most villagers were concerned, this more or less fully explained how women and men divided their duties. Very soon, however, it became clear to me that transgressions of this basic labour division regularly occurred. In other words, women could often be seen performing some 'big' labour and it happened that men engaged themselves in 'small' things, even if this occurred much less frequently. Nonetheless, there was a compelling consensus considering this general definition and division of male and female farm labour.

The purport of 'big' work included a variety of activities. Practical chores like carrying heavy things, for instance manure and rice sheaves, the harvested rice, cut and tied into bundles by women, as well as other duties where actual muscle power was needed, such as pulling a cart full of manure, was obviously 'big' work. Indeed all the transportation involving carts pulled by cows and buffalos fell into the masculine 'big' sphere. However, managing and decision-making responsibilities, such as expanding or reducing the number of fields, trying out a new type of rice seed, or planning what other crops than rice to grow, were also part of men's 'big' work. The two 'biggest' activities in terms of actual farm duties were ploughing and harrowing. At least ideally, the male farmers were the ones who worked with ploughs and harrows, together with their cows or buffalos. As mentioned already, women spend more time than men labouring the fields; but from a male perspective, this did not necessarily imply that women were capable farmers in their own right. Several male farmers argued that they lay the ground, so to say, for women to work on – something they literally did when ploughing and harrowing the rice fields before women transplanted the rice.

Ideas about gender-based farm labour division were strongly intertwined with other relations associated with power or control between adult men and women. For instance, both men and women commonly argued that men were better suited for 'deep thinking' and managing larger projects, like building houses. This meant that they should have the final say in important decisions concerning, for example, economic investments beyond everyday purchases, repairing the house, digging a fishpond, or organising weddings. Men were generally perceived to possess greater analytical abilities than women. In the patrilineal family, men were supposed to be the centre-pillars (*cot tru*) of the household. Men's good decisions were expected to be the base that women needed for performing their everyday duties. 'Men are supposed to be leaders, like countries have leaders', my landlord told me one evening in a rather grand manner. 'Women are women and men are men, and a man should be the director. He should know how to do everything in theory, but he does not do it all in practice. His duty is to organise, lead and manage the work of the household', he explained to me.

Women's 'small' farm work included chores that were perceived as light, easy, and delicate, i.e. work that did not require great muscle strength but, on the other hand, required the assumed feminine capabilities of endurance and patience. Women's ability to work swiftly with their hands, working in a bending position for long periods of time, and being careful and painstaking in their work were acclaimed attributes of femininity. Transplanting rice seedlings into the water-filled fields was by far the farming activity most associated with women. This duty required what was understood as particular female abilities of dexterity, swiftness and bodily suppleness. The 'smallness' of female work also implied that women's chores required less mental efforts and were of minor importance. An example is when Nam, the leader of the Women's Union in the commune, argued that women tended to concern themselves with everyday matters of making daily purchases, cooking, cleaning, doing laundry, and caring for children, in addition to their farming chores. Nam implied that they did not think so much, and often felt insecure in matters dealing with things other than housework. Men, on the other hand, were more preoccupied by things like how to improve land use in order to increase the income from farming and they acted more like managers (see also Gammeltoft 1999). Wolf reported very similar ways of reasoning in relation to general female and male capacities among Chinese peasants who, just like the Vietnamese, have been under the influence of socialism for decades without really revolutionising gender roles and relations (Wolf 1985: 84–85). In many of the households in Lang Xanh, Nam's claim held some veracity, but to my understanding, there were several exceptions to be found. However, this was a salient discourse concerning gender capacities and relations and, as in Nam's assertion, it gained support from more official directions as well. Han, the commune's cadastral officer also agreed that men managed the farmland, while women spent more time actually working on the fields. He lived according to his opinion. During my time in the village I often met his wife Huong while she was working the fields of their household, but never did I run into him doing any kind of farm work. Han spent most of his time indoors where he, among other things, kept check on the administrative aspects of the village fields.

Young recently married women, who were still novices as farming wives in their new households, were gradually incorporated into the feminine sphere of 'small' work. Phong, a very young man who had recently married an even younger woman and now shared a household with his elderly and unhealthy father, revealed to me his strategy for transforming his still relatively inexperienced wife into a capable woman. By learning from her mother, his wife already knew how to do many things in theory; he believed that now was her time to practice actually doing them. The things she did not really master very well yet were for example, sowing the seedlings for transplantation or raising pigs. As far as he was concerned, he might know how to do these things better than her at the moment, but he let her do them for the purpose of gaining practical knowledge. Phong came from a farming family with relatively limited means. When he was a boy, he had helped out where ever he was needed so, like many young men in Lang Xanh, his knowledge about 'small' work was not so small. However, as a married man, he did not actually intend to perform many of the activities that were meant to be carried out by his wife. Instead, he executed his power as the 'supervisor' of his young wife and let her be moulded by her assigned tasks into a competent female farmer.

Thus, despite the fact that the men in Lang Xanh normally spent considerably less time working in the fields, their tasks were in many situations attributed with more prominence compared to women's chores. These kinds of general representations of gender relations were commonly interjected in everyday discourses and formed a grid through which the work of men and women was perceived. Using the statement 'women are like children', or to take an example from Lang Xanh 'women do "small" work', Moore argues that, the often broad and general, yet compelling, dominant representations of certain gender characteristics become powerful because they are habitually used in everyday life and thus set standards for what is appropriate (Moore 1996: 177–78). Dominant gender perceptions provide people with standings and orientations towards themselves and other people in daily situations and even when contested, these representations of men and women form a basis from which conduct and responsiveness emanate. Similar to Lang Xanh, Moore noticed from her own fieldwork among the Marakwet in Kenya, that the actual time and effort that women invested in performing all of their farm labour duties was not reflected in the dominating discourses on the commitments of men and women in agriculture (ibid.: 177).

In line with this basic idea about gendered labour division in Lang Xanh, some of the farmers emphasised the economic advantages attached to the principal order of work. It would be a waste of male as well as female capacities if men were too occupied by 'small' work and women tried to carry out 'big' things; the entire household would suffer from such an ineffective distribution of responsibilities. It was for the benefit of everyone if each household member concentrated on what they did best, and this was achieved when men and women acted in accordance with their 'natural' capacities. A transgression of the gendered labour division could be acceptable if the circumstances of a household had been temporarily altered, due to unforeseen matters like sickness or, as will be discussed later on, the shorter or longer absence of household members of working age. In practice, farmers' individual qualities also influenced the division of tasks. However, a well-functioning farming household consisted ideally of husband and wife at the centre, with adolescent children (if they did not study) and members of the older generation as extra labour resources. When a household suffered a lack of female farm labourers this seemed, in many ways, to be more problematic than when there was a shortage of male labourers. We will return to this later in this chapter.

Femininity at work

The intense workloads of contemporary Lang Xanh women were integrated into their feminine identity in numerous ways. One example was The International Women's Day (8th of March), which was paid a great deal of attention in Lang Xanh, as it is all over Vietnam. The cadres of the commune gathered at the People's Committee and several speeches were delivered to address and honour Vietnamese women. A popular topic, which was introduced at the one-year anniversary of being appointed a 'cultural village', concerned the hardworking and courageous efforts performed by women during the wars. Traditional female virtues such as dignity, tolerance and sweetness were praised and women were reminded never to fail to care for their children and encourage their husbands in their work. The local women were appreciated for their ability to raise children who were well cared for, build up 'cultural' households[1] and engage in agricultural labour. The increase of divorces (there were two ongoing cases in the village) was mentioned as a criticism and as a warning signal of the threatening decline in morality, which less diligent women might cause. This combination of praise and admonition was significant for the occasions when the officially proclaimed epitome of Vietnamese women was articulated in Lang Xanh. Women could always perfect themselves, even if they already worked hard (see also Drummond 2004).

It was not only on the 8th of March that women's glorified war efforts set the context for discourses concerning the hard work that women were

1 Households without conflicts and where the members did not engage in any 'social evils' such as gambling, drinking, drug-use or promiscuous behaviour, but instead showed diligence by hard work and good studying were referred to as 'cultural' households (see also Drummond 2004).

capable of – and still today were expected to perform. When Vietnam was at war against France (c. 1945–54), and later on with the USA (c. 1960–75), many men had been absent from Lang Xanh as soldiers spread all over the country and into the neighbouring countries of Laos and Cambodia. This forced women to carry the main responsibility for the sustenance of life in the village, in the families and as farmers. 'During the war [against the USA], when the village consisted mainly of women, children and elderly or sick men, women had to do everything', Viet, a 54 years old man who had then been a soldier, told me. In a discussion about male and female farmers, Lap, a former party cadre and one of the oldest men in the village, considered that the local women had been exceptionally self-sacrificing during the war against France. According to him, they had mended and washed the soldiers' uniforms and they had sung for the wounded soldiers to send them to sleep. 'People used to say that women had a plough in one hand and a rifle in the other hand. As part of the resistance against the French, women were given the so-called *three responsibilities*', said Lap continuing his story. These included the responsibility to care for state affairs, for the household and for fighting. 'Women's responsibilities were resumed in the war with USA, when women were left in the village for ten, maybe even 20 years without any news of their husbands', he added. Our conversation had started with a discussion on the farm work of men and women. Lap, however, quite soon moved into a more ideological track and gave an example from the war, in order to convey the attitude that he felt characterised the assiduous work of Vietnamese women. Good women, as Lap wanted to convey to me, remained faithful and hardworking also during tough times; this was a kind of woman who was worthy of respect. Stories like this regularly came up, in for example official speeches at meetings and celebrations, and worked to direct women's attention towards an ideologically informed idea of their labour. In addition, like in the situation with Lap, it was a way to present an ideal version of Vietnamese women to 'outsiders' (see also Nguyen-vo Thu-huong 2005).

When men's admiration for women's hard work was voiced it was usually phrased in an air of splendour. It quite often happened, as in the example above, that Lang Xanh men started to praise Vietnamese women as a response to my inquiries about women's farm work, but they rarely celebrated particular women. Instead, they commonly talked about women in general and their ability to sacrifice themselves for the country and their families, without any concern for their own well-being. One man put it like this: 'Women get up early in the morning and go to bed late, always working for the sake of their husbands and children and always giving them the best food, while settling with inferior food for themselves'.

Female capability – how to accomplish every task

'It is hard to be a Vietnamese woman' was a comment that reached my ears numerous times, almost every day. Remarks of this kind were usually made by women in reference to how arduous their farming duties were, and how much more work that awaited them when they returned home from the fields. Regardless of the norms of labour division in which men were supposed to do the heavy and 'big' work, many women found themselves doing much of the work belonging to men as well. Some women ploughed their fields and one woman, as mentioned, offered her ploughing services to other households. Women also engaged in a great amount of hoeing, mostly in the manioc fields, but sometimes in rice fields. They even carried heavy loads such as manure, when no man was available or ready to do it. It was a common belief among women that husbands were not always so dedicated to their farm duties as their wives could have wished for, something that also has been noted in other studies of rural areas in Vietnam (Endres 1999: 158; Gammeltoft 1999: 163) and in China (Wolf 1985: 107). An example of this arose during harvesting, when I met a small group of women managing the rice sheaves without any men present. Men often carried the rice from the fields, but these women told me that the men in their families, if they had been present at all, were already back at their homes having a cup of tea, smoking, or watching TV. Other women recurrently conveyed similar observations. It also happened quite frequently that the male members of the family, as the representatives of the households, attended a wedding, a funeral, or some other event where drinking was involved. This could make the men indisposed for the rest of the day and then women had to put in some extra hours of labour to compensate for the absent men. Women's responsibility for the 'small' everyday duties included the ability to accomplish virtually all household and farm work if this was needed for one reason or the other.

Only the more bold and outspoken women (qualities that were often enhanced with advancing age) explicitly said that they thought men were more inclined to avoid hard farm work than women were – 'men' being seen as a general category, or as their kin or neighbours. They also said that men tended to show a careless and negligent attitude to the many tasks that formed part of their daily lives as farmers. Statements like these were commonly made in informal situations where only women were present who knew each other and felt comfortable with each other. However, the great majority of women did not hesitate to claim that they, in their own right, were very able farmers. A common phrase that women often used was that they '*biet lam het*.' This means to 'know how to work until it is finished.' I interpret this statement to imply that women not only knew how to fully complete an agricultural task, but also have the capacity to carry out any farming duty. In other words, most women were confident that they could manage cultivating the fields on their own, without much help from men. Hong, the leader of the village Women's Union predicted that soon women might actually do half of all the ploughing that was required in the village – the most 'male' farming chore of them all. When men were not fit or available, women had to cope on their own, she concluded. Hong herself had two buffalos and she was the woman in Lang Xanh who was engaged by other villagers to do ploughing and harrowing. She managed the farming of her household, while her husband worked as a construction worker in another province and only came back on a handful of annual occasions.

Lang Xanh women quite frequently suggested that men, in contrast to women, did not possess the capacity to make everyday life run smoothly. Men did not cope well with the continuous making and remaking of the home place. This was a female competence that men depended on women to accomplish and it formed a necessary complement to men's roles as overall 'supervisors'. When it came to farm work, I frequently met women who either giggled or grunted with some contempt at certain men's self-proclaimed roles as farming supervisors. A couple of the women simply stated that 'men do not know how to grow rice.'

In the very beginning of my stay in Lang Xanh, two female friends of my landlady came to visit our house one evening after dinner, to chat and watch some TV. It was in the middle of the winter transplanting and they had spent the day in the cold water of the rice fields. They started out by telling me some quite ghastly stories about leeches creeping up women's legs and into their bodies while they worked in the watery fields.

The two women observed me closely to see my reaction to these stories[2] noting that I certainly found them unpleasant. But I was also curious to know why there were no men present in the fields at transplanting. The women replied, quickly and in unison, that men do not know how to transplant. They giggled as they added that men only know how to do 'night-time transplanting'. Their own bluntness then made them slightly embarrassed as they realised that they had just made a joke of obviously sexual character even though we did not know each other very well yet. This was not, however, the last time I heard jokes about the sexual implications of a gendered division of farm work.

Transplanting was the domain of women and their area of expertise. Hence, alongside the undervaluation and subordination of women's 'small' work and the hardships of their workloads, came a sense of pride and confidence where many women gained and showed strength in their ability to cultivate the land. The female chores might be less prestigious but they were not insignificant and they were certainly not small in number. Women often expressed the view that they persisted in their work despite demanding circumstances. This was perceived to be a female competence that could be facilitated by – but was not entirely dependent on – the presence of men.

Cultivating female morality

In Lang Xanh, the farmers argued that to get a good crop of rice it was important to spend adequate time in the fields. The previous chapter showed that prayers and offerings to otherworldly beings were not perceived to improve the chances of a good harvest. On the other hand, critical factors for farming included sufficient labour in combination with knowledge and fine weather. The more time that was spent in the fields and the more attention they were given, the better the outcome would be. If a farmer seemed to have too much idle time, other villagers commonly saw this as a sign of carelessness and bad moral behaviour. Besides, the fields were there visible for the public eye and for everyone to judge. Everybody knew which field belonged to which household, and one quick glance was enough for the villagers to decide if sufficient labour had been employed.

2 Similar stories came to my knowledge throughout the fieldwork and I suspect that they might have strains of urban legends in them.

On one occasion, Lan – who was my landlady and well known for her diligent farm work – informed me that she did not intend to participate in the singing performance that was to be held in the village. 'You do not earn any money from singing', she stated rhetorically. To emphasise her point, Lan conveyed to me her ideas about the moral state of the woman living across the street. She and her husband were going through a divorce and as far as Lan were concerned, that woman had spent too much time on leisure and too little on work, while relying on her husband's apparently faltering ability to earn money. Now they were separating and had to sell their house. Lan, for her part, preferred to concentrate on the well-being of her family and what little time and energy she had left was not to be spent on entertainment outside the home. She told me that people had tried to persuade her several times to participate in singing events in the village or in the commune, but she normally refused, often by referring to all the farm work she had to do. It was important for Lan to devote her time to her family and labour. Respectability and high moral standing were for her incompatible with being seen out amusing herself in public. As far as Lan was concerned, a fun-loving woman with some weed in her rice fields possesses very little virtue.

From people's comments I could tell that laziness and indolence were regarded with great disapproval in general, and this was especially true in the case of women. As we have seen, women sometimes thought of men as lazy (*luoi*) when they had to perform farm work. This was perceived as annoying by women, but also something they just had to accept as a rather common male characteristic. Female laziness, on the other hand, was not accepted in any way and women were quick to criticize other women for being lax and neglectful. Lan, for instance, pointed to an empty rice field and remarked that no corn was planted in between rice crops on that field. The woman who tended this land plot was obviously lazy and this would prevent her from gaining any extra income. Other women's insufficiencies were frequently used to shed some light on one's own good traits. Women closely monitored each other's behaviour. This was clearly illustrated by Lan and her teenage daughter during a break from work after lunch one hot summer afternoon. All of us were resting and the two of them had been lying in bed watching TV. Suddenly we heard the dogs bark as a sign that somebody was coming. In a blink, Lan and her daughter jumped out of bed and ran into the kitchen, pretending to be occupied with some household chores. They had suspected that the visitors might be a couple of older women dropping in on Lan's mother-in-law, who shared the same household as Lan and her husband. This would not have been a good occasion to lie idle, because older women's judgments of a daughter-in-law could be severe, even when she had reached her forties.

A handful of Lang Xanh women of working age were perceived, by themselves and by others, as 'weak' and therefore did not do much farming. One of them ran a small shop selling daily necessities, like washing detergent, matches and sweets, together with her husband. She spent much of her time in the shop while her husband did more farm labour than most of the other men in the village. Many of these 'weak' women often stayed at home, in or around the house, and were seldom seen out in the village. As Gammeltoft has showed in her work on Vietnamese women and their experiences of illness and stress, physical weakness could be a legitimate way to convey other kinds of worries and distress, which if expressed in other ways could be interpreted as self-absorption or lack of proper virtue (Gammeltoft 2001: 277). However, in the case of the female farmers in Lang Xanh, this approach was not always successful – especially because, as pointed out, women scrutinised each other closely for moral flaws. The weakness of some of these women could not be derived from any easily recognized diseases or from other obvious signs of poor health, nor did it seem to be due to other kinds of suffering. Because of the lack of apparent reasons for the incapacity of these women, the sympathy was not always great and it was quite common to hear remarks about them being lazy rather than ill. It was suspected that they let their husbands do much of the farm work that women were supposed to perform. Women who carried out their work despite difficult circumstances, such as physical pain and bodily weaknesses, were on the contrary held in highest esteem (see also Gammeltoft 1999). Women who could be suspected of trying to avoid their share of labour took great moral risks.

If we again recall the *four virtues*[3] derived from Confucianism, then labour (*cong*) is the first virtue. Ngo Thi Ngan Binh, in her article *The Confucian Four Feminine Virtues,* notes that according to tradition, the

3 The virtues are: 1) appropriate labour, 2) appropriate appearance, 3) appropriate speech, and 4) appropriate behaviour.

work of Vietnamese women should not only be accomplished, it should also be performed 'in the proper manner and at the proper time' (Ngo Thi Ngan Binh 2004: 64). From her research on daughters-in-law and mothers-in-law in contemporary Ho Chi Minh City she found this to still hold true, especially among the women in the latter category. In order for a daughter-in-law to be regarded as mastering the fourth virtue, which is good behaviour/character (*hanh*), it was of great importance that she executed the first virtue, labour, with utmost care and diligence. *Hanh* could only be achieved by executing the daily chores properly – in attitude as well as in actual skill. A young woman's devotion to her family-in-law was shown by the manner in which she performed her everyday work. By engaging in self-cultivation, she could, over time, build up her character and gain respect as a morally strong woman (ibid.: 64–65). The connection between women's work and their relation to, and position in, the patrilineal family was reflected in Lang Xanh women's farm work, as the following section shall explore.

Women's farm labour and the patrilineal family

Cong was a man in his 50s with two young daughters and a wife who was a farmer. He had been injured in the war against USA and due to his pains, he could frequently be found at home during the daytime. He often invited me for tea, offering interesting conversation. On one of these occasions, he told me that it was of utmost importance for the state of the family whether or not a woman knew how to organise her work well. According to him, the wife was responsible for carrying out most of the work of the household, which did not preclude that the husband sometimes assisted her with things like cooking and laundry. He told me that preferably a wife should know how to share her work with her husband, and most importantly, how to encourage him to work. A man, he continued his reasoning, might do his fair share of the work, but without encouragement he might do it in anger and without much care. In other words, women's skill and attitude towards work was essential for the atmosphere of a family. As a father of two daughters that would eventually become wives, he intended to teach them how to be gentle, hardworking and encouraging towards their husbands. According to Cong, women's hard work was crucial for keeping the husbands in a benign mood and for supporting their will to work. This, in turn, meant that the family would hold together. If a woman failed to handle the work situation well, quarrels that might eventually lead to divorce could easily occur. In such situations, it was often the wife that suffered the most, because she was usually blamed for her inability to create a good atmosphere for the husband and children to dwell in; she might be accused of failing to compensate for her husband's hotter temper.

Cong's ideas about women and labour reflected the commonly held opinion that if a wife behaved properly and possessed proper feminine virtues, family discords did not need to arise or grow large. This idea was also expressed in the village meetings when the topic of divorce came up, as mentioned earlier. Vietnamese women should be able to keep up their daily work and endure to preserve family harmony (see also Endres 1999; Gammeltoft 1999, 2001, Pham Van Bich 1999; Rydström 2003; Schuler et al. 2006). In a regional perspective, Wolf noted that in rural China, when she inquired about the kind of characteristics that signified a 'good' husband and a 'good' wife, the ability of wives to be hard working topped the list. The same answer was given both by men and women. For husbands, an even temper was the most desirable trait as far as women were concerned. This was even said to compensate somewhat for lack of helpfulness (Wolf 1985: 234–35).

Consequently, what ultimately made the work of Lang Xanh women rewarding in terms of virtue and morals was performing their work for the benefit of their families rather than as an individual pursuit. When a married woman engaged herself in farm work and took pride in keeping the fields in a good condition, she also cultivated her care and respect for her children, her husband, and her parents-in-law. Lan, as we saw, was careful to point out that she avoided the singing events in order to use the time to work for her family's welfare. The self-sacrificing actions of labour and endurance that well-performed farm work required were tangible proof of female loyalty. Women commonly remarked that they endured the efforts of farm work first and foremost in order to care for their children. Later on, the children would hopefully take care of their elderly parents. Consequently, a dominating moral belief in Lang Xanh was the idea that women who did not, through persistent work, put their families as first priority could never have a harmonious family situation.

A woman who found herself in a precarious position due to her ongoing divorce was Hanh, who was in her mid 30s and, at the time, lived with her two small sons in her parents' house. She had entertained the idea of taking up a more lucrative job than farming. Applying to be a maid in a city, or even abroad, or to do business buying and selling things in markets bigger than the local ones had passed through her mind. However, the temporary market activities she had engaged in, as a married woman, had been one of the things that had triggered the divorce. Her husband thought that she cared more about going off to markets and doing business than she cared about her family. According to him, Hanh had changed her personality when she engaged more extensively in market activities outside the village. She became more demanding and started to criticise him in ways she never did before. Now Hanh was afraid that if she left the village to do other work than farming, her husband would take the two sons and let them be raised by his mother. She, however, was careful to point out that all she wanted was to work hard to make sure that her sons were cared for; that was the only thing that mattered for her. In Hanh's present position, she felt that her safest choice was to stay and work on her fields. Any other actions would have negative influences on her moral standing.

Factors like limited access to farmland and hopes for an improved economy had made some women in Lang Xanh pursue other lines of work than farming. A handful of married women in their 30s had, like Hanh, travelled to more distant markets where they had done business for a few weeks at a time, when farming did not demand their presence. They claimed that Lang Xanh women in general were too timid and lacked the courage to go away to buy and sell things at markets, even though it could be quite lucrative. Other women, however, considered that to be able to do this, you needed a defective consciousness and an unsophisticated personality. A handful of younger, or at least still unmarried, women had left the village to be market traders or street vendors in Hanoi, or to work at restaurants cooking for construction or factory workers. One young woman had left to work in a factory in Malaysia[4] and another one had left for China to work with something that was rather unclear. Not even her parents were sure what she did, but it might have something to do with being an assistant and interpreter for a Chinese woman. Most of these women were surrounded by rumours

4 For accounts on young women working in factories, see for example Thu-houng Nguyen-vo, Tran Ngoc Angie and Mandy Thoman in Lisa Drummond and Helle Rydström (eds), *Gender Practices in Contemporary Vietnam*, NIAS Press, 2004.

and gossip about them being prostitutes and those who had recently left were constantly reminded by their parents not to get into trouble. Some of the parents ensured that they trusted their daughters, but they also considered that young women were always more vulnerable than men. If it was at all possible, parents tried to send their adolescent children to live with, or near, a relative to make sure that they were safe. As far as women's labour was concerned, the only place outside the home that did not put women's morality and virtue at risk was the fields. Virtually all activities outside the physical, as well as moral, sphere of farm labour within the village held great potentials for transgressions of moral demarcations.

Men's esteem and their responsibility for the family unit were less precarious, not least due to men's firm anchorage in the patrilineal kinship system. The temporal mode of women's belonging to their husbands' families – as explored in the previous chapter – tied female morality closely to their daily work inside the patrilineal family. Lang Xanh women were required to constantly work on their virtues as respectable members of the patrilineal family. The fields provided a proper place for this labour to be performed. As noted above, a daughter-in-law's way into her husband's family and her way to gain their respect, was achieved through her carefully performed labour (Ngo Thi Ngan Binh 2004). In addition, practices such as patrilocality contributed to the necessity of women to work out a place for themselves as 'outsiders' brought into their husbands' patrilineages by marriage. This was discussed at greater length in chapter 3. Men, on the other hand, were not constantly obliged to display their belonging and devotion to their family; to some extent this was taken for granted, since men in themselves constituted the patrilineal family.

From a certain perspective, it is possible to understand women's unlamented and selfless work and endurance, which contributed to the preservation of the patrilineal family, as female passivity and subordination. As the family positions of Lang Xanh women often were of inferior character, their relatively lesser status provided them with little room to question or to resist their duties. This was particularly the case for younger females who were not yet mothers-in-law. Hoan, a young woman, pregnant with her first child lived with her husband in the same house as her parents-in-law. For some time now, her husband was engaged in some temporary work in Hanoi, while she stayed and worked as a farmer. Hoan ascertained that it was the fate of women, and especially of daughters-in-law to work really hard. Before her marriage, when still living with her parents, she had not had to work this much. Now she found herself to be the main labourer of her new household. While her mother-in-law assisted her in some of her chores, she was expected to carry out most of the farm work. If she did not do this, she ran the risk of upsetting her husband and her parents-in-law. She suggested that she might even be beaten if she neglected her chores – a potential threat that hang over many young daughters-in-law. Dignity and virtue were best preserved by silent compliance. Talking back to a husband or a parent-in-law could only lead to trouble in the family and was regarded as lack of respect and filial duty – two definite signs of immorality. Therefore, the frequent absence of open resistance to their situation was at the same time women's way to safeguard their position in the family. This kind of 'active' compliance included elements of hard work, because, as Gammeltoft notes, 'endurance does not come naturally, but requires a lot of effort in everyday life' (Gammeltoft 1999: 213).

Since women's work was so closely bound up with respectability, nobody else could really compensate for a woman's absence or evaded work efforts. If, or when, men performed 'women's labour, the meaning and implication of the work was altered, as the following section shall explore.

Masculinity and farm work

When women's work expanded into the male area of 'big' things, this did not necessarily pose a threat to the basic order of labour division. It happened frequently that women carried out 'manly' work, when they managed most of the farming by themselves. The dominant gender discourse surrounding the sphere of farm work was still left intact. As we have seen, women's performance of 'big' things could contribute to a woman's respectability and good morals and men could still retain their positions as 'directors'. Men's relation to farm labour was, on the other hand, more frail. Even if the majority of the men in Lang Xanh were to some degree engaged in farming as well, their relation to farm work was more constricted as far as transgressions of a gendered division of labour was concerned. Men's farm duties were fewer and they risked, rather than improved, their status and reputation when they deviated from what was perceived as ideal behaviour. Men who ventured into the realm of 'female' farm duties had little or nothing to gain from this in terms of prestige or morals. Male positions in Lang Xanh were manifested as much by the farm work that they *refrained* from doing as by the labour that they actually performed.

At an early stage of my stay in Lang Xanh it became apparent to me that most men felt awkward doing womanly things like harvesting and, to an even greater extent, transplanting. Many of the villagers knew of men who used to transplant, but it was always talked about in past tense, declared as a consequence of exceptional circumstances (e.g. poverty and lack of female labourers), or done by someone who lived in another village. When men talked about male transplanters, they always gave other men as examples and very rarely admitted to this kind of activity themselves. Nevertheless, a handful of local men were pointed out as more or less experienced in transplanting. Most commonly, people mentioned two twin-brothers, now in their late 30s, who came from a poor family where the mother had been a widow for a long time. Their household had lacked labourers when they were young and the two brothers had been obliged to work hard to assist their mother in any task that was needed. When I asked them about their experiences of transplanting they were not too keen on the topic and simply declared that they had not had much of a choice. Someone had to do it, and lacking women or the money to hire female labourers, the task of transplanting had fallen to them. However, this had been part of their childhood, now they were both married and their wives did all the transplanting that was needed.

The only two men I saw transplanting rice throughout my stay in Lang Xanh were the adolescent sons of a divorced and unwell woman. She also had an older but still unmarried daughter – one of the street vending women who, according to hearsay, had a dubious reputation – who did her best to return for a few days at the peak of rice harvest or transplanting. But for a few busy days of summer transplanting, her two sons also helped. Their situation resembled the twin brothers when they were young. Like the rest of the very few men who admitted to transplanting, these two sets of brothers stressed that if they, for some reason, had to engage in this activity, they would never do it on somebody else's fields. They would only do this kind of labour as an exception to the rule, when their families really needed it. In contrast, women quite often transplanted other people's rice fields, as part of their labour exchange; or in some cases, for payment, as will be dealt with in the next chapter. It was obvious that transplanting, with its close association with femininity, did not attribute male actors with any desired qualities.

Risking and protecting masculinity

While they were busy with transplanting the summer rice crop, Huong and Thuy, two female farmers in their mid forties, recounted their opinion of men's attitude to farm work. As far as they were concerned, it was often inadequate. Thuy told me that her husband went out drinking rather than working on the fields. Even if they only had vegetables to eat, he did not care. In the adjoining field, Huong was transplanting with some other women while one of her teenage sons carried the bundled rice. The son laughingly suggested that if his father – Han, the cadastral officer in the commune – would go to work on one of their fields, the sky would open up and it would immediately start to rain. This was his way of telling me what an earth-shaking deed this would be. On the one hand, this kind of behaviour was generally understood to be part of men's 'natural' inclination to be 'lazy' (*luoi*) and somewhat defiant – male traits that women had to endure. However, laziness as a male characteristic existed predominantly as a concept in a female context. From a male viewpoint, men had to be careful in their farm work in order to avoid crossing the fine line between being hardworking and helpful and appearing to be submissive or effeminate. In this sense, refraining from women's 'small' work was often a matter of cultivating appropriate masculinity. Nonetheless, a number of women spoke warmly about their supportive husbands, and to a certain extent, assisting one's wife in her farm work could add appreciation to an already respected man. Practiced in excess though, it could harm a man's image. Nevertheless, some men did engage in undoubtedly female duties like harvesting or, as the above examples showed, even rice transplanting. In these cases the men were running the risk of being apprehended as unmanly and '*so vo*' (afraid [of the] wife).

'*So vo*' was a frequently occurring concept and a popular topic for jokes in Lang Xanh. Everybody knew somebody who they thought was 'afraid of his wife', who acted obediently and submissively towards her.

For instance, one elderly woman complained that her son, who lived in a neighbouring village, never came to help her with any farming. This was because he was '*so vo*', and his wife kept him from coming to assist her. Men who seemed to accept being bossed around by their wives could easily fall victims to gossip and condescending remarks. Cong, the man with the two young daughters, told me that he was hot tempered, and a man of few words, while his wife was more talkative and sociable; but he did not batter his wife, even if their characters were quite different. The reason why he did not beat her was not because he was afraid of her, but because he respected her, and this, he carefully pointed out, was an important difference. There was no equivalent concept for women who were afraid of their husbands, Cong made clear in his reply to my questions. According to him, marriages where wives were scared of their husbands and silently abided their will, were not unusual, or a deviation from convention.

Even though the concept of '*so vo*' was often used as a joke, it certainly had serious connotations. A man who was not the 'manager' of the family was perceived to lack respectability. From this perspective, it was important for men to avoid performing duties that, due to their female character, should preferably have been performed by their wives or other female family members. This became evident at harvest time, for example. Then it was possible to observe men of all ages – but most commonly younger ones – who preferred to wait in the shade of some trees for the rice bundles to pile up before they started to take them back from the fields, arguing that they, unlike women, did not know how to cut the rice well. By claiming their ignorance of a feminine farming duty, these men engaged in masculine self-fashioning. The young men expected to be heads of households, if they were not already so; hence their status was not of the kind that would benefit from performing female farm chores. It was more important for them to show that they could afford to keep to men's work.

A number of Lang Xanh men in their late forties or older suffered from some kind of war injuries, even if their wounds were not always immediately visible, as was the case with Cong. Therefore, they were unable to do farm work, and many other kinds of work, without constraint. These men were referred to as being weak (*yeu*) and many of them spent much of their time at home where they did such work as caring for the animals or cooking – tasks that women normally did while their husbands worked outside the home. These men had to struggle not only with their war injuries, but also with the predicament of not being able to contribute much labour to their households. The wives of these ex-soldiers had to perform many heavy duties. Either the women themselves had to do such work as hoeing or carrying manure, or another man had to be employed to do it. Yet, in most cases, their past as soldiers fighting for the country compensated for their present insufficiency.

Even so, these former soldiers had to guard their masculinity when performing what little farm work they could do. At times, they joined their wives on the fields, although not for the most feminine duties. Most commonly, they helped on fields that held crops other than rice, because female farming skills, as pointed to already, were predominantly associated with rice cultivation (e.g. Le Thi 2001: 2–3; Le Thi Nham Tuyet 2002: 18). When I met these men working alongside their wives, they were often quick to point out that they did it mainly 'for fun' (*cho vui*), as a voluntary pastime that they could have chosen to spend in another way – watching TV or chatting with a neighbour for example – but had now decided to pass like this. Another way, used by men as well as women, to account for why men sometimes could be found labouring together with their wives on the fields was that they had taken some time off from their own work to help (*giup*) the women with their duties. The emphasis on helpfulness allowed men to distance themselves from some of the farm work they were doing, while still being able to acknowledge the fact that they were engaged in 'small' work that should have been performed by women. By maintaining the voluntary and temporal nature of their engagement in female farm labour, men could sustain the distance in prestige between 'big' and 'small' work.

If men refrained from doing their 'big' farming chores, it was important that they did it for accepted reasons. Renouncing 'big' farm work was best tolerated when it was replaced by another 'manly' activity, like building, carpeting, factory work, or holding an official position. Men who deviated from this and lacked an accepted reason were also referred to as weak, but their weakness was seen as a flaw in their character rather than a physical state induced by external circumstances. For instance, the first time I met the father of the local barber – the latter a married man in his 30s who was also a musician at funerals – he presented this son as being a weak (*yeu*) man. The father had been a farmer all his life, but his son had never done any farm work, not helped out or even played at being a farmer as a boy, he told me. He had been weak from the beginning, not cut out for 'big' work. Cutting hair and playing a traditional musical instrument did not so easily fit into his father's ideas about masculinity.

When a married man ventured outside the realm of appropriate male farm labour, the respectability and morality of both him and his wife were at risk. This was demonstrated by the reasoning of Lan and Minh, the married couple that I lived with. One day, a man told me: 'Your landlord is good at transplanting, but he would never tell you about this.' The idea that Minh should be skilled in transplanting rice somewhat surprised me, because he seemed quite sensitive about male and female work. He had, for instance, stopped helping his wife with the laundry when I moved in because he was embarrassed to be seen doing a female duty. When I asked him about transplanting, my landlord explained that he had not really transplanted by placing rice seedlings into the muddy grounds of the fields, like women commonly did. He had used the method of broadcasting the rice seeds straight into the fields, like a sower. Later on, I learnt to better understand why this was an important difference. However, he did confirm that he had done quite a bit of farming when the children were small. Nowadays his wife did all of it. Together with Lan, he explained that because he was a full time schoolteacher and their children were almost all grown-up, it would seem strange if he had to work as a farmer as well. Most certainly, his wife assured me, his colleagues at school would tease him if he laboured the fields. They might get the impression that she in fact gave the orders around the house rather than him. Moreover, his colleagues might think that Lan was not capable of organising and coping with her farm work. In their case, Minh's engagement in farm labour could hurt both their reputations.

It was Giang and Quy, two women in their forties, who eventually pointed me towards another dimension of the embarrassment that was attached to the idea of men doing 'women's' work. Giang admitted that she was not very skilled at transplanting because her household had the benefit of having quite a lot of labourers when she was a girl. Therefore, she used to concentrate on her studies rather than doing farm work. Eventually she married a farmer and she had to work on the fields even though she lacked the proper skills. However, she would feel very sorry for her husband if he had to transplant. She said it would be embarrassing (*nguong*) for both of them. At that point of our conversation, her neighbour Quy, who was one of the most efficient transplanters in the village (many times, I had admired her swift hands when she handled the seedlings) came by and joined us for a while on her way to the fields. She claimed that men might know how to transplant, but the embarrassment of pointing their bottoms towards the sky made them avoid this task to the utmost extent. Men did things that they could do in an upright position, like harrowing, ploughing and carrying things, she explained. As soon as she said this, I realised that I had not seen many men working in a bending position in the fields. A couple of days later Quy shared a proverb with me in the middle of transplanting a field. A man says to his wife: 'poor you, you always have to stick your bottom up towards the sky' and the woman replies: 'you should not say so, because I alone have to bend down, but it feeds your whole family.' At this point it was clear to me that the embarrassment did not only concern men's risk to be conceived of as weak and 'afraid of their wives', it was also a matter of bodily awkwardness and shame, when men bent forward to do a feminine task like transplanting rice.

Harvesting was another task that required working in a bending position and very few men did this. Tuan, married to one of the 'weak' women who did comparatively little farming and instead spent most of the time working in their little grocery shop in the village, got some first hand experience of the 'risks' involved if a man harvested rice. One morning in the very early harvesting season, he was 'caught' cutting rice together with two women. These women worked to pay off some debts they owed in the shop and he had come to harvest with them on his own. Around the field were two men grazing their buffalos and they took the opportunity to chat with the harvesters while the animals were feeding on the freshly cut rice stalks. Squatting on the banks of the field, the two men started to tease Tuan, saying that judging from the work he did, he must be half man and half woman, because no real man would bend down to cut the rice and bundle it up (see also Gammeltoft 1999: 178). All three of them laughed and since Tuan was not really arguing against their statement about his mixed-up sex, they concluded that their assumptions must be correct. 'Probably', one of them continued, 'he also likes to lie underneath'; i.e. he most likely took the position of a woman during sex. This was another body position that was not intended for men, but was reserved for supposedly passive and receptive women.

In her study among the semi-nomadic Moores of the Sahara desert, Popenoe noted the analogy between men's mobility and female immobility when she saw that men in their most virile years covered geographically vast areas, while women at the height of their fertility spent much of their time in their tents, hardly moving at all. This was also reflected in ideals about the sexual act, where it was stressed that men should be active and always positioned on top of an immobile and receiving woman (Popenoe 2004: 161). I noted similar discourses in Lang Xanh, where men were not only expected to do 'big' things and be 'on top', but also presumably be interested in 'tasting' some *pho* (noodle soup) when they leave home, as a change from the *com* (cooked rice) they normally had at home. *Pho* is here a synonym for a mistress or an alluring woman who entices a man by offering something different from the safe and well-known home sphere where *com* – i.e. the wife – is the staple food. There is also an analogy between geographical and sexual mobility, where men are allowed to be the active 'movers'.

Men left in charge of women's farm work

If a Lang Xanh farming woman left the village for a longer period of time, the extent of her husband's predicament when it came to 'female' farm labour became even clearer. The same year as I arrived in the village, a handful of women started to contemplate the possibility of going to Taiwan to work as maids. In Hanoi, several agencies that arranged for Vietnamese women to go abroad to work as housekeepers had opened. According to rumours, there was good money to be earned and the required labour was supposed to be less back-breaking than farming. Eventually a small group from the commune left, among them were two women from Lang Xanh. Both of them were married; one had a small son and the other had two teenage girls. When the two of them went to Taiwan, their husbands were left to manage the households and the fields without their wives for a long period of time. A couple of the women in Lang Xanh followed the event with the absent wives with interest and they calmly informed me that these two men would now have to hire women to do all the farming duties which their wives normally did. As far as husbandry was concerned, they continued, few men with absent wives normally kept anything more than a few chickens and maybe one pig to cover for the most pressing needs. It turned out that these women's prediction was very accurate.

When the woman with the teenage daughters first introduced the idea of going to Taiwan to work as a maid, her husband did not seem too excited, but when his wife persisted, he changed his approach to the issue. In between farming, he used to work temporarily as a construction worker; this he claimed gave quite a good income. Now he began to argue that since his wife did not really earn much income from her farming anyway, she might as well leave and make some money for a change – he would manage just fine while she was away. Proudly he announced to me that he could do all the farm work his wife used to do, and he could do it much quicker and better. When she finally left, he lent most of their fields to his younger brother's family. The few rice fields he kept were transplanted by his two teenage daughters and with the help of some female relatives and friends. After the rice was harvested, when virtually all of the farmers planted corn on the fields as an intercrop, he decided not to cultivate any corn that year. However, with some help from his daughters, he did plant manioc on the hill fields. This was quite often done by men and women together, even if women on average spent more time on manioc planting than men did. In preparation for the next rice crop, he also pre-cultivated the rice seedlings that were to be transplanted on his fields. This task was not specifically 'female' either, even if more women than men did it.

The other man – the one with the small son – did not do any farming whatsoever when his wife left. He also let out or lent most of the fields to related households in the vicinity and he had female relatives to do all the transplanting and harvesting on the few fields he kept. His old mother took care of his son while he was away doing business at different markets in the area; and for the next rice crop, he did not grow any rice at all.

Naturalised gender division of farm work

The powerful and seemingly universal gender order of 'big' and 'small' work derived much of its persuasive character from its ubiquitous presences in a myriad of situations in people's lives, and especially in the context of farm work. A gender-based labour division was perceived to appear by its own power, without any detectable source, and was shared

– even if not agreed upon – by everyone (Bourdieu 1990: 69–70; Moore 1996: 177–79). The different skills that male and female farmers possessed seemed, according to the predominant view, to be immanent in the bodies and minds of the performers.

In Lang Xanh, men as well as women explained to me that to be a farmer who worked the land – more or less without any mechanical equipment, as they did – was the most natural way of living a life. They did not oppose, or go against the innate capabilities of men and women; instead, they ordered their lives in accordance with nature. If we remember Viet, the male farmer whom I quoted in the beginning of the introduction, he put it like this: 'the land tells us what to do and all we have to do is follow. We feed the land and the land feeds us.' A couple of men with whom I discussed this matter were convinced that if they had machines at their disposal to do many of the tasks, which they now did manually, things might very well look different. As one of them shrewdly pointed out 'anyone can press a button', implying that the less labour the farmers did by hand, the greater were the chances of reducing the differences between the work of men and women. Presumably, machines would even out the gendered differences by performing the jobs that normally set the sexes apart due to their inherent differences in character. Male and female farmers shared the notions that the outcome of rice cultivation by means of manual labour reflected the gendered capacities of the workers. Machines, on the other hand, were regarded as working independently from the operator. It did not matter if the fingers that triggered the functions of the machine belonged to a male or a female hand.

Women often argued that men's bones are harder and their bodies are stiffer. In addition, men are clumsy with their hands and lack the patience needed for more delicate work, like transplanting the rice seedlings. This literally made men quite unsuitable for some of the tasks involved in rice cultivation. Men more or less agreed with this perception of the male body. They confirmed that transplanting made their backs hurt after only a short time out in the fields and those men who did not have much experience from this particular chore assumed that they would quickly suffer pain and stiffness if they ever gave it a try. Due to the lack of swiftness in their hands, men transplanted slowly and distributed the rice seedlings in an uneven fashion on the fields. Women's bodies, on the contrary, were naturally well suited for these types of farm skills.

Dam and Kha, two elderly women with a lifetime of farming experience, argued that in many cases the fairly well-known male proneness to laziness was a reason for men's reluctance to do women's work. However, when it came to rice transplanting, male indolence was ruled out as an explanation. The two women were convinced that men refrained from this task due to bodily differences. It was men's hard hands and tall backs that prevented them from handling and planting the frail rice seedlings in a skilful way.

The order of a gendered division of agricultural labour was often explained by the farmers as being simply 'given' (*danh cho*) to them or, alternatively, it was decided by 'god' (*Troi*)[5] (c.f. Bourdieu 1990: 71). In other words, the principal division of labour between men and women appeared to be natural and self-evident, without any detectable reasons or origins. It would be to go against sense and nature to purposely alter this order. The naturalised labour division was sometimes illustrated by referring to the old and popular proverb, stating that 'the husband ploughs, the wife transplants, and the buffalo harrows.' Farmers explained that as long as they keep working by hand, the trio of man, woman, and buffalo would be at the core of their work – each of them performing the skills they mastered best. This saying was presented by the farmers in Lang Xanh as a symbolical and almost primordial structure of local gender relations. On several occasions, this division of labour was even voiced in terms of gender equality (see also Le Thi Nham Tuyet 2002: 35). In these cases, the farmers argued that when men, women, and nature cooperated and shared their work based on a division of labour where land and people were used in an optimal way, then equality – or rather equilibrium – must reign. According to the farmers in Lang Xanh, the meaning of gender equality – which had been introduced by the Communist Party as a desirable goal – could not be to create disorder and ineffectiveness in farm labour by changing the positions of men and women. I think that it is possible to sense influences from several lines of thought that had merged together in the farmers' perception of a 'natural' division of labour. For one, there seem to be traces of a communist, state-proclaimed ideal of non-hierarchy and equality, but also of the rationality and efficiency that were supposed to govern the citizens' labour in an advancing society. The

5 *Troi* literally means Heaven but is also often referring to a celestial sovereign divinity, (c.f. Nguyen Van Huyen, 1995: 245).

connections between a well-functioning society and the citizens' 'given' roles and positions can also be derived from the Confucian notion of a fundamental order of things, where for instance women and men had their given and complementary – but not interchangeable – positions to fill (see also chapter 1).

In Confucian cosmology, gender traits and relations are perceived as an integral part of a universal order. Male and female are linked to the complementary opposition of *yin* and *yang* (*am/duong* in Vietnamese), which ultimately relates men to women in the same way as the Heaven relates to the Earth. Harmony is reached when each person occupies her or his proper position, and acts accordingly (Gammeltoft 1999: 173–74; Rydström 2004: 75–77). In contrast to communist ideology, Confucian ontology is not based on equality but on an ordered hierarchy where each person respects her or his position in the social structure. As pointed to in the introduction, a common perception among many Vietnamese is that the human body is essentially male or female, with adherent gender-specific traits and functions. This view includes an idea of a relational position of men and women that is contextually decided and complementary but, at least ideally, not interchangeable (Rydström 2003: 40, 92). This is evident in how femaleness is associated with *am* (the Chinese *yin*), earth, passivity, cold, subordination, and the negative; while maleness is connected with *duong* (the Chinese *yang*), heaven, movement, hot, superiority, and the positive (Rydström 2004: 75–76). At the same time as this incorporates men and women into a mutual and dynamic universal order, the gendered bodies of male and female persons set women and men in different positions towards each other and the world.

The Confucian conception of the relation between men and women is accompanied by definitions of how other relations should be, such as between older and younger persons; superior and subordinate; citizen and the state. Together, all these structured and hierarchical relations become an ontological principle that – in its generality – is an all-encompassing and compelling system of order and meaning (c.f. Bourdieu 1977: 93–94; Moore 1996: 178–81). As many of the villagers believed, gendered capacities are an integral part of how the world is constituted and labour is 'naturally' divided in accordance with this worldview. Taken together, the farmers' perceptions of the inherent capacities and inseparable interaction of bodies and land that was 'given' to them, maybe by the 'heaven', or by 'god' (*Troi*), formed a primordial ontological order.

Persisting 'natural' labour division

Gradually, I realised that the division of farm work in Lang Xanh had been imbued by the idea of female 'small' work and male 'big' work probably since before the 1945 revolution. The state's shortcomings in enhancing gender equality in farm labour were evident in the experiences that farmers in Lang Xanh had from the time of collective farming (c. 1959–89). According to some women who I discussed this with, the collective's working brigades included both men and women, but the activities were largely divided into male and female chores. There was a tendency to give women all kinds of duties, but virtually no men did any 'female' chores. This resembled the situation in Chinese agricultural collectives, where women's labour tended to expand into more traditional male areas, while men's labour assignments stayed relatively true to what used to be typically male tasks (e.g. Croll 1995; Jacka 1997). These trends have been noted in socialist countries on a general level; commonly the quest for gender equality has been eclipsed by an idealisation of the solidly established female roles as mothers and caregivers, while established male roles have never truly been called into question (see also chapter 1). In most socialist states, the demands and expectations of even greater labour efforts on behalf of the state were added to women's already comprehensive domestic duties. These demands were not complemented by the state propagating for men's expansion into 'female' domains of work (Molyneux 1981: 178, 197–98; see also Mai Thi Tu and Le Thi Nham Tuyet 1978).

Thus, during the height of collective labour, women's work was, at least partly, redirected from the 'traditional' family towards the state. Within farming, the traditionally female tasks could, especially during warfare, expand to include virtually any labour duty. Women had to step into the male realm of labour when the political, economic or family situation so demanded, but it was rarely expected – from official, or from private and family domains – that men crossed the borders of gendered work. Neither women, nor men in Lang Xanh could recall that men did any of the 'female' chores during collective farming; as far as possible, the gendered division of labour was maintained. If there were any transgressions, they were made by women. Even so, the largely unaltered household chores that women performed within the household retained the female workers in the farming cooperative firmly in their conventional positions in their homes and families. In this way, household work functioned as a strong counterbalance to the labour that women performed in the public, more 'male', sphere of the cooperative. We can conclude that as far as men were concerned, the farming cooperatives never posed a real challenge to established ideals of masculinity, or gender relations for that matter. Even if men had been deprived of their land-owning rights and much of the power to organise their families' labour (see chapter 3), the duties they performed as cooperative labourers had still guaranteed them a certain preservation of established masculinity because they had continued to perform predominantly 'big' tasks. By and large, the farming collectives had divided the chores of farming between women and men in the same way as 'heaven', or 'god', had 'given' them in the past.

Farming skills and the making of gender

An important part of the 'naturalisation' of a gendered labour division in farming was the passing of skills from one generation to another. For instance, young females were incorporated into rice transplanting by joining older female relatives in the fields. None of the women in Lang Xanh ever considered teaching their sons to transplant, this was strictly saved for the daughters. The skill of ploughing followed a similar pattern among men. Male as well as female farmers presented these kinds of skills as something that was not taught in the more strict sense of the word. Young people do not 'study' (*hoc*) farming chores, as one father of two young sons told me, they just learn through being there on the fields, by playing and experimenting at a young age and then, little by little, they develop a sense for the task at hand. By gradually acquiring particular farming skills, the young and inexperienced bodies of girls and boys incorporate the history and tradition as they experience it being embodied by the more experienced farmers (Ahmed 2006: 41; Bourdieu 2000: 154). The fields were a location where the shaping and orientation of girls and boys took place. They encountered the chores, the equipment and the environment that over time would comprise and form a familiar world; a place where it appeared 'natural' for women and men to direct their attention to particular things. Using the weaving of string bags among Telefol women in central New Guinea as an example, Ingold presents the idea that skill is reproduced from generation to generation by giving novice bag weavers the same task to practice over and over again. This results in a process of enskilment (Ingold 2000: 357). Ingold's way of reasoning about skill resembles how Jackson regards our bodily movements and practices to be inseparable from our ideas and culture. Jackson argues that bodily performances are not the result of preconceived ideas, but *are* – due to their practical nature – reality in a direct way. 'Persons actively body forth the world; their bodies are not passively shaped by or made to fit the world's purposes.' (Jackson 1989: 136). In other words, persons, their activities and the things they use merge into habitual places.

If we take skill to be a way of modifying and tuning the body by repeatedly practicing certain actions in specific places – like transplanting rice fields – then a gendered division of labour skills would imply that male and female bodies are tuned in diverse ways according to the skills they learn and incorporate as young actors. The world they 'body forth' – the opportunities for movements and actions that open up in certain places – would be characterised by the gender of the body which orients itself towards the world (Ahmed 2006: 46, 60; Young 2005). Without making a point of it, Ingold notices that Telefol girls start to develop their skills as basket weavers at the same time as their bodies start to form into adult women. The skill, he argues, is therefore growing into the body rather than attached to it. The swiftness and flow of weaving become an intrinsic part of being a Telefol woman, especially because virtually all women acquire this skill – just like rice transplanting for Lang Xanh women. Telefol men do not weave baskets. Their hands maintain the clumsiness of novice makers (Ingold 2000: 360) – much like the men in Lang Xanh when they try to transplant. What Ingold exemplifies is not only how skill is acquired but also – and this is a crucial aspects that he leaves entirely unnoticed – how girls are made into proper young women by embodying certain knowledge. To embody a gendered skill is, thus, to acquire a body that is modified in a specific way because of its repeated engagement (in the world) in a mode specific to one's gender. In this way, skilled actors also become gendered actors. Hence, the bodies of men and women in Lang Xanh had become the kind of masculine and feminine bodies that were either stiff and hard with slower and more fumbling fingers, or weaker and suppler, with swift and dexterous hands. In his elaborations on the nature of skill, Ingold holds that acquiring a skill involves incorporating perceptions and actions to the extent that they virtually become innate to the actor (Ingold 2000: 358, 360). We can then further conclude that acquiring gender-particular skills is a way of becoming a person orienting oneself to one's surroundings in accordance with one's gender (Ahmed 2006: 60). When the women of Lang Xanh performed their 'feminine' farm duties and for instance, spent time bent forward transplanting a rice field – they became directed to the world, to activities and places, as females.

Lang Xanh women's femininity, manifested by traits like selflessness, endurance, dexterity, and sacrifice for their children and families, could not be separated from the fields and their farm work. Rather, these traits were cultivated every time a woman worked a field. At these times, such feminine characteristics were not abstract ideas about ideal women, but were the actual worldly circumstances and lived reality of a Lang Xanh woman. Alongside women's actual presence on the fields, the work, skills and body postures they performed during farm work informed them about their 'fate', their restrictions, and their possibilities as women. Returning again to Ingold's ideas on skill, he holds forth that by repeating bodily practices in particular environments, the actor develops a 'feel' for the task at hand (Ingold 2000: 352–53). However, when Lang Xanh women learned and performed the skill of transplanting rice, I believe that it is possible to expand the acquired 'feel' for the task also to include a development of a 'feel' for which kinds of femininity that could be enacted and cultivated in this particular place.

We can further say that women and men's work opened up for, and made certain tasks and places available for each other. Men, as they themselves argued, prepared the ground for women to cultivate rice in the fields. They literally did this when they ploughed the fields, also by taking on an overall responsibility for larger household decisions they, as they said, 'gave' (*danh cho*) most of the actual farm labour to women. However, by doing most of the farming and the absolutely largest share of the everyday household work, women opened up the field of 'managing' and 'directing' the household for the men. Women's work gave men time to think, reflect and discuss with other men (Ahmed 2006: 31).

From this perspective, women's 'feminine' and 'small' farming duties were a direct realisation of the status, responsibilities and possibilities that women held in their families, in the fields, and in society. When women engaged in their agricultural tasks they simultaneously sensed how they, according to their gendered position, could cultivate themselves and what was required of them in order to become respectable female persons.

In the case of men, as we have seen, only a limited number of farm chores could be practiced without making themselves vulnerable to accusations of unmanly conducts. They got the 'feel' for their positions as leaders of families and society and as permanent manifestations of the patrilineages by standing up straight and being strong and self-determinate, i.e. by restricting their farm labour to include primarily 'big' things.

CHAPTER 5

Cultivating community – gender and labour organisation

'[W]ork cannot be measured separately from relationships', Strathern (1988: 160) points out in her study on Melanesian society. Following along that line, this chapter will show how relations based on farm labour and exchange of farm work also involved the forming, maintaining, and expression of kinship and community ties. For Lang Xanh – a village where most of the households were still involved with farming – whom to work with and at what task were vital issues for the nature of intra-village relationships. Another major issue was how to handle compensation for labour, when this was needed. Cooperation and compensation were areas in which community-oriented values like solidarity, unity, and equality had the potential to be enacted in the shape of farm work. Farm labour also held the potential to imply differences in equality and independence between the households, and an increasing dissociation from village community through an increased use of paid labour.

As shall be highlighted in this chapter, the predicament of either weakening or preserving a sense of a place-oriented village community is to some extent tackled by the complementary roles that women and men play in the organisation of labour (e.g. Carsten 1989; Moore 1988; Strathern 1988). This dilemma had to be addressed by the farmers in Lang Xanh every time they worked their fields. The way in which male and female farmers organised their labour was a statement of their attitude to their work, as well as to their neighbours, friends, and relatives. Farmers had to balance their actions and decisions carefully (c.f. Scott 1976: 42). This chapter shall show that these balancing acts were different and involved diverging moral impacts on male and female farmers.

A first illustration of how entangled farm labour could be with other social relations is provided by the following episode. In mid-May it was time to plough the fields in preparation for the second rice crop of the year, and Hanh, the mother of two sons who now lived with her parents while her divorce was progressing, was contemplating how to cope with this. As previously mentioned, she was in the middle of the torments of a divorce and this had implications for her managing and organising farm work. Normally, her husband would have been the obvious candidate for ploughing the fields, especially since he was more or less a fulltime farmer and the owner of a buffalo as a draught-animal. But because of the separation she did not intend to cooperate with her soon-to-be former husband. They had divided their fields between them and were now organising their farm labour separately. Hanh had already tried to hire several ploughmen, but so far nobody had agreed to undertake the job. No one wanted to become involved in the ongoing dispute between her and her husband, not even for money.

Unfortunately for Hanh she did not have any male relatives from her natal family who could plough her fields. And she could not ask Hong, the leader of the Women's Union in the village, who was the only woman who ploughed other people's rice fields for payment, because Hong's husband's father was the brother of Hanh's husband's grandfather. Even if Hong was only distantly related to her husband, it was not suitable to ask her to do the ploughing – neither for hire nor for exchange of labour.

Ploughing was a specialised 'male' skill that many villagers paid for, but in this case, when relations were tense, the social aspects overshadowed the possibilities of a more business-like agreement. The divorce proceedings meant that even a hired ploughman risked having his moral standards called into question. He could be accused of ignoring moral aspects in favour of monetary benefits and be a greedy man who 'died for money'. He might also be suspected of having an affair with Hanh and therefore willing to compromise his moral conscience. It was unthinkable for Hanh to ask a man to exchange labour with her, for example, by offering to transplant rice for him in return. This would be regarded as a too intimate and lasting arrangement between a not yet divorced woman and an unrelated ploughman. Eventually, Hanh's field was ploughed by a man who neither she nor her husband was related to, and on top the husband claimed to be paying for the service.

Despite the intentions of the state, farming in Lang Xanh was not a fully commercialised activity, performed according to the mechanisms of market economy. It was rather part of an all-encompassing way of life where money, self-interest, communality, and morals worked together governing daily life and social practices. The organisation of farm labour were in many ways revealing of a household's priorities, needs, and values, and placed the farmers and their families in various relationships of commonality and hierarchy upon which their social security, repute, and sustenance relied (c.f. Humphrey 2002: 167–70; Scott 1976).

Working together

In many respects, matters related to the cultivation of the land united virtually everyone in Lang Xanh. People shared ideas and concerns about farming when they met in their homes, in the village, and out in the fields. Especially rice cultivation demanded a certain level of cooperation. The field holders had to be in agreement about the methods for and timing of certain stages in growing rice. For instance, some types of rice seeds did not grow properly when planted too close to other types, and if transplantation was not properly coordinated, the crops could suffer from attacks of insects and rats. If, for example, one farmer transplanted a field much earlier than her field neighbours, her crop ran a greater risk of being damaged by vermin. Likewise, at harvest time, a farmer who was behind schedule risked loosing much of the remaining rice to the inevitable rats.

Anh, one of the war invalids who engaged in some agricultural tasks to the best of his ability, asserted that if you do not cooperate and communicate with the persons who farm the fields adjacent to yours, you cannot count on them showing much consideration for your fields. When people do not cooperate, farming becomes difficult and ineffective. Irrigating the rice fields – something that was normally done by digging a hole in the surrounding bank on the side facing the water flow, and then patching it up when the water level was right – was an activity that needed to be made in cooperation with neighbouring field holders. When a rice field was bound together with fields belonging to other farmers, which was almost always the case, it was necessary to lead water through their rice fields in order to irrigate your own. If this was done without consideration for the neighbours, Anh explained, it could happen that newly applied fertiliser was washed away and ended up in someone else's field. Because the fields were so closely connected, all major activities, from irrigation to transplanting and harvesting needed to be performed in coordination with the surrounding field holders.

Besides the more general concern and communication, there were certain periods during the farming year when the farmers needed to form working relations with people outside of their own households, in order to cope with peak seasons. This was particularly common during rice harvest and transplantation, and in connection with ploughing and harrowing to prepare the fields for the next crop. During these periods many farmers needed the same type of labour at more or less the same time. Each household tried to recruit labourers into working teams in order to accomplish these major tasks as quickly and timely as possible. For the greater part of the year, however, farming did not take up an unreasonable amount of time. Most of the farmers distributed the labour duties first and foremost among the members of their households. With three or four able labourers available, a household only needed extra workers at the peak periods of the farming year.

Farm labour recruitment

My first inquiries about how the villagers organised their farm labour, and whom they usually worked with, normally produced the same answer: 'We work with anyone.' Especially on 'public' or first encounters, male as well as female farmers usually expressed an including attitude, where kin, friends, and neighbours engaged in reciprocal exchange of farm labour, and 'anyone' who were available when work was needed would be included in the farming community. 'We ask the first person we meet if that person can come and work' it was commonly explained to me. Such ideas of intra-village inclusion and equality corresponded well with an idealised idea of communality and village sentiment/spirit (*tinh cam*),[1] and were also in accordance with both Confucian and socialist ideals, where mutual assistance rather than individual accomplishments was given preference (Abrami and Henaff 2004: 113). But these comments also sprung from the farmers' knowledge about the norms and conditions that governed the recruitment of co-workers. They knew, of course, who could be asked to do what, where, and with whom. And

1 *Tinh cam* translates into feeling, sentiment, or emotion and can refer to a personal capacity or to a shared feeling, like in village sentiment.

because they knew this, any one of the suitable candidates could be asked to cooperate and participate in that particular farming task.

The female domination of farm work meant that women – in spite of the general idea of men as 'managers' – performed most of the organisation and recruitment of labour in the households. They did this by drawing on their networks of consanguine and affined relatives and non-kin. As the previous chapter showed, when a household lacked women of working age, men usually let their young daughters and other female relatives organise the work, or hired other women to perform the female farming tasks. Very rarely did male farmers recruit others into a working party on the fields of their household, and it was even more unusual if they themselves joined in such a group. For women, this was part of their working routine. Daughters commonly engaged female relatives from their fathers' and mothers' kin groups and supplemented with some friends and neighbours. When married, women recruited other women from their natal kin groups together with women from their husbands' families. Women married to men with relatives living in the area, could also include male kin and friends of their husbands' in their labour groups, provided that these men were to perform the more 'male' tasks, like carrying harvested rice. In this way, the organisation of farm labour in Lang Xanh entailed the formation of extensive networks and communities of women who cultivated the fields together.

During transplanting and harvesting periods, the fields were filled with women who worked together, most of the time bent over the rice that was to be either planted or cut. Most of them wore conical hats and some had covered their faces with scarves or textile masks, as protection against wind and weather. When I had lived in the village for a time, a few of the women liked to put me to a test by pointing to the women working in the field, one by one, and ask who they were. This was not an easy task. However, my ability to recognise particular features of individual women improved and sometimes I passed the test without too many mistakes. Besides speeding up my ability to recognise the villagers, these tests also gave me a sense of the composition of the groups, which reflected the field holder's social, economic, and kin circumstances, as well as her personal preferences when it came to gather together a group of labourers.

Many women found it easier to make flexible and convenient agreements with kin. When working with relatives – in-laws as well as natal kin – the chances of relatively smooth cooperation were usually greatest, many thought. The mutual and sympathetic aspects of kin relations were then emphasised. My landlady Lan, for example, often exchanged labour with one of her sisters-in-law who lived not very far away. She explained that: 'Working with relatives is easiest because you can be more straightforward and you can agree to stay a bit longer on a field in order to finish it even if it is getting late.' Her experience was that the working days were more carefully counted and regulated when non-kin were involved. And as nearly every household in Lang Xanh was related to one or more of the other village households, most of the farmers could choose to either give precedence to family or to neighbours and friends.

I found Dinh, a married woman approaching her 50s and mother of two teenagers, a girl and a boy, sitting by a patch of land where rice seedlings were grown for later transplantation. She uprooted the little plants and tied them into handy bundles to be brought to the fields. Arguing along a similar line as Lan, she told me that she usually worked only with her two sisters 'because it is much less complicated'. She explained:

> With others you have to negotiate more and make sure that everything is fair, and it could be that you have to pay in cash if you cannot find the time to pay back in labour. Or you might have to ask somebody else to do it [i.e. pay back in labour] for you and it becomes complicated. It is easier with relatives and you can coordinate your work all the way.

Considering the small-scale and household-based nature of farming in Lang Xanh, close and adaptable working relations corresponded well with many of the households' labour needs. Cooperating with relatives, where the emphasis was on social relations and mutuality – rather than more commercial aspects where labour efforts were measured according to efficiency and profitability – suited many of the farmers well. From Lan and Dinh's point of view, establishing working relations with unrelated households was normally imbued with more careful considerations regarding equality and fairness, and the time for reciprocation was normally shorter.

Nonetheless, close neighbours were also an appreciated category of co-workers. For example, many of the unrelated women who worked together on the fields in high seasons explained that they had been good neighbours for a long time and they enjoyed and valued dearly the time they spent in each other's company. Moreover, some women pointed out, through their frequent cooperation they eventually picked up the same working pace as their co-workers, something that enhanced the much valued socialising that went on at the same time. If a family had moved from another part of the village, or from a village nearby, they quite often kept the working relation with their old neighbours who, over time, had turned into good friends.

There was, of course, a multitude of reasons why some villagers avoided relatives as their main co-workers. Conflicts and tense family relations were often, but not always, the reason why labour agreements with neighbours or friends, rather than relatives were preferred. Farmers who preferred to work with villagers who were not their relatives found support for their behaviour in the concept of communality and mutual assistance (*tinh cam*) between fellow neighbours and farmers (c.f. Luong 1992; Scott 1976). The female farmer Nhung, for example, found it easier to ask friends and neighbours, instead of adapting to her four sisters' labour schedules. She thought that this gave her more freedom of choice when it came to decisions about when and with whom she wanted to work. With relatives, the level of constraining obligations could be experienced as outweighing the empathetic aspects. The village's only divorced male farmer Vinh, for example, shared the opinion of Nhung. He too preferred other villagers rather than relatives for labour exchange. He saved his relatives for special family-oriented occasions like death anniversaries and funerals, events when the mutual assistance of relatives was inevitable. Irrespective of how the villagers got on with their relatives, the idea of intra-village community and 'we work with anyone' presented a complement to – or, rather, coexisted with – kin-based unity and support.

Emotional attachments between farming neighbours, as a result of their spatial closeness and cooperation in farm work, was not at all new to the village. In fact, this had been part of the sense of a shared community before, during, and after the collective period and had resulted in long-term and close relationships. This was, for instance, expressed in a popular saying that I often heard when the villagers spoke about cooperation: 'selling distant relatives and buying close neighbours' (*ban anh em xa mua lang gieng gan*). As this proverb illustrates, friendly relations with those living close by was regarded as more important than remote relatives, whom you hardly ever saw or did not know well.

Ideals and ideas of the rural village

Notions of community sentiment and spirit (*tinh cam*) have for a long time been vital ingredients for the image of the 'traditional' northern Vietnamese village. This conception is a significant part of the idea of 'home place' (*que*), that holds a central position in the lives of urban as well as rural dwellers (see Schlecker 2005). One of the first questions that strangers asked when meeting each other for the first time was: 'Where is your home place (*que*)?' Most rural and urban inhabitants have a village in the countryside that they count as their place of origin. This can be the place where they were born or grew up, but it can also be the village where their parents, grandparents or ancestors came from. The idea of the *que* is surrounded by an air of romanticism, in the sense that the inhabitants of the rural village are pictured as honest, helpful, and empathetic. As opposed to the city, where people are believed to be strangers to each other and do not care for or help other people, the village is assumed to be the essence of community spirit where you can always count on your neighbour (Drummond and Thomas 2003; Schlecker 2005).

As an effect of the economic renovation, as well as the globalised economy, Vietnam has seen a surge of nostalgia for an idealised image of the rural village. Urban life is, in this respect, regarded as emotionally empty, harshly materialistic, and a threat to genuine Vietnamese values. Uncertainty in times of instability, economic difficulties, and an increased urbanisation, has paved the way for a longing for another way of life, mainly among the more affluent city dwellers. This course of events has been underpinned by the government's focus on 'cultural' issues as a means to curb social fragmentation, and endorse a notion of national community. According to this discourse, the genuine Vietnamese values – such as communality, care and consideration for others – that should be defended against 'outside' corruption have their roots in the country's 'traditional' rural villages. The perceived obsession with money and profit by the urbanites, in contrast to the empathetic and caring ways of the rural villagers, is not a new phenomenon in Vietnam, but has been reinforced by the shift to a market oriented economy (Drummond and Thomas 2003: 7–9; Schlecker 2005: 512; see also Nguyen-vo Thu-huong 2005: 3).

The villagers of Lang Xanh were well aware of this discourse and sometimes made use of it when reflecting on their situation. Van, a farmer in his 50s, told me:

> We might not be rich in terms of material standard, but we are rich in emotion (*tinh cam*). Few people hire labour because exchange is better for the emotions. In towns and cities it is different, but here we have very little influence from technology.

This was said while he was contemplating the great amount of low-tech manual labour that the farmers engaged in. What he considered as outdated and labour-intensive methods, were to some extent compensated for by the community spirit that labour cooperation helped to maintain. From this perspective, the attitude of mutuality and exchange of farm labour between the villagers, that Van expressed, existed in the same 'natural' sphere as the gendered labour division, which was also perceived to be in line with manual farm work, unaffected by technology and modernity. He explained that the villagers did not always keep a very close track of the working days that they spent helping other villagers with farm chores. The idea of assisting each other without calculating the expenses and seeking personal gain, was a constituting part of many, if not all, of the villagers' self-perception. Embedded in this ideal was the notion of villagers who supported community-based solidarity, and wanted to share their work and give their support to ensure that all households had a certain level of sustenance (see also Scott 1976; Taylor 2007).

From the perspective of villagers like the farmer Van, the solidarity that the inhabitants of Lang Xanh preferably showed towards each other was contested by the growing use of paid farm labour. Some farmers feared that when certain people became wealthier, they did not want, or need, to cooperate and socialise with the others to the same extent as before. Moral obligations to assist each other and exchange farm labour to cover the seasonal peaks used to be the basis of village community, they claimed, and in order to hold on to this, the people in Lang Xanh needed to continue to contribute their time, their energy, and some of their means. If the individual quest for money overshadowed the will to engage in communal work, the local community risked destabilisation (Douglas 2003). Van, for example, believed that this was already the case in the cities, but could still be avoided in Lang Xanh if the villagers continued to involve themselves in farm labour exchange and mutual help. A sense of village spirit also demanded that the villagers were in place to practice the sharing of activities. Monetary contributions – for labour or other activities – were not sufficient to make a community; this also required actual presence and involvement. Douglas argues: 'When a large crowd turns up, community solidarity is made visible' (ibid.: 147). From Van's viewpoint, farmers who did not involve themselves in relationships of labour exchange with fellow villagers demonstrated their disengagement from the local community and tangibly contributed to the dissolution of the same.

Agreements on labour compensation – a moral balancing act

Side by side with ideals of village communality lived the notion that people of an advanced and modern society should be able to work and gain an income, without taking into consideration relationships based on accumulated obligations to kin and friends and the maintenance of a place-based village communality (Malarney 2002: 211; Schlecker 2005: 521). With the introduction of the economic reform (*Doi moi*) in 1986, the government has propagated private and commercial farming (Abrami and Henaff 2004: 95–96; Taylor 2007). Vietnam's entrance into the global market economy made the glow of labour exchange somewhat fade, due to its association with small-scale and subsistence-based farming with rather low efficiency (Le Cao Doan 1995: 122–23). This relatively new politico-economic situation gave rise to some moral re-evaluations. For many of the farmers in Lang Xanh, access to money signified material success and provided the option of hiring labour, and organise work in a way that released people from the time-consuming task of recompensing other people's labour efforts by an equivalent amount of working hours. The time gained could instead be used for some more lucrative activity. However, it was first and foremost the more affluent farmers who could use paid labour as a means of rationalisation and increasing profits. But as Van pointed out above, when a household used money to buy labour, it became less dependent on local cooperation so that the equalising effects of labour exchange weakened, together with the sense of a shared community.

To a certain extent, the sensitivity which surrounded paid farm labour could also be related to the prevailing socialist ideology – including the firmly established aversion against landlords who did not work themselves, but paid others to work in their fields (see chapter 1) – which had governed farming for approximately forty years but came to its end with the 1993 land allocation. However, even though market economy and individual land use rights had been introduced, some of the villagers in Lang Xanh were nonetheless motivated by socialist ideals. The Marxist-Leninist ideas, which had formed an ideological base when the state introduced collective farming, asserted that capitalism causes a shift in society, from communality and mutuality, to individualism and self-interest (Kerkvliet 2005: 138–9; Malarney 2002: 51). In other words, the moral bonds that knit people together in communities of solidarity are destroyed by the calculating and impersonal capacities of money (Parry and Bloch 1989: 4, 6). From the perspective of the state, farming had undergone an ideological change; the socialist objective of communality had been substituted by an individualistic economy based upon entrepreneurial households. For the farmers in Lang Xanh, though, farming encompassed sustenance as well as social and moral relations and positioning.

When the farmers tried to organise their labour, they also had to consider the moral implications of their decisions and practices. The frailty of the place-based village community, the need for cooperation and assistance, and the requirements for more 'effective' and profitable farming methods all had to be addressed. Each choice and action that both female and male farmers made, represented a kind of positioning in relation to the village farming community and, notably, it made them susceptible to evaluations of their moral standards by the other villagers. As the following sections will explore, women and men had to take these aspects into consideration – however in different ways – when they operated within the field of labour organisation and compensation for farm work.

Reaching an agreement

When farmers, male as well as female, recruited labour from other households, it varied from one time to another whether the workers were to be paid back in work (i.e. labour exchange) or in money. In general, it was the relationship and requirements of the involved parties, rather than the type of labour, that determined if it was money or exchange of labour that was an appropriate compensation. It happened that the same persons sometimes exchanged labour and at other times compensated each other in cash. Depending on the parties involved, as we shall see below, any kind of compensation was not suitable for any kind of work. Factors like gender, kin, and friendship played decisive roles when it was decided if the compensation should be in cash or labour, or of it was as situation of unremunerated help. The circumstances of each worker as well as of the field holder were taken into consideration when making these agreements. Within a group of people working together on a field, each person had his or her own agreement with the farmer in charge of the field.

Among people who exchanged labour, the general rule was to count working days. Commonly people added up the days within one working season, and then regulated the balance between them at the end. If some discrepancies between two persons had arisen during the period, it could be evened out by cash or, more commonly, by keeping the issue in mind until the next working season. Good friends claimed that they trusted each other and normally did not feel obliged to settle debts immediately and, as we have seen, among relatives the calculating of working hours was not always done very carefully. In the busiest periods, farmers had to prioritise whom to exchange labour with, because there was not enough time to do this with all the partners involved. The result could be that the co-workers with the most distant relationships to the field holder had to be compensated in cash instead of time. It was the closeness of the relationship, as relatives or friends, which usually decided the order of precedence. The obligation of repaying each other in working hours was strong among the most affectionate friends and the closest and dearest relatives. Monetary compensation was unthinkable if a relationship was considered to be warm and friendly – intimacy was manifested in the time people were prepared to give each other (see also Bourdieu 1972: 180; 1990: 115; Carsten 1989: 123).

A certain degree of calculating did, however, often take place within the realm of farm labour, even when kin and close friends were involved. Unless a household needed help due to sickness or other unfortunate circumstances, people expected that the time and labour effort that was contributed should eventually be returned in equivalent amounts. To perform at a level below one's ability was not acceptable, and the capacity of each household was closely monitored by kin, friends, and neighbours, and formed the basis for expectations of labour contribution.

Female farmers and labour compensation

When female farmers engaged in farm labour exchange, it could for instance look like this: Vy, a woman in her late forties, harvested rice together with four other women in her field. Two of them were her neighbours and the other two her mother-in-law and her sister-in-law. They were all exchanging labour no monetary compensation was involved. The neighbouring women were persons with whom she had formed a trusting relationship over time, and who largely shared her circumstances as a farming woman. As for her relatives it was really more a matter of helping each other, as she put it. Over time, they would give each other a hand to the best of their abilities. Particularly elderly parents were assisted with little expectation of any kind of compensation. This was a way for children to show filial piety towards their parents in the realm of farming. Among close kin, it was a personal responsibility to be present and to provide labour and this could not be passed over to just anybody, or substituted by paid labour. In this sense, when close, commonly female, relatives assisted each other with farm labour they also cultivated their affiliation. The same principles were applicable for close friends and neighbours too, but normally on a more voluntary and equal basis and sometimes with a stronger emphasis on fairness. The social and moral significance would be lost if monetary remuneration was involved.

Therefore, the question of monetary compensation was critical part of farm labour organisation. Paid labour was a topic that was treated with wariness by all parties involved. This was, for instance, reflected in the fact that people rarely spoke of hired farm labour by its real name, but rather used the same modes of expressions as in labour exchange by either using social and kin related idioms – or, as often happened, people just said that they helped (*giup*) each other. Despite the fact that almost everybody in a working team knew whether their co-workers were exchanging labour with the field holder or were paid in cash, people preferred not to talk about this in the group. In a private conversation with a farmer, the nature of compensation of the labourers could be discussed, but not out on the field and not in the presence of other villagers. When I began to enquire into the issue of labour compensation, upon meeting groups of women in the fields, my questions were sidestepped with jokes or met by awkward silences. Nobody wanted to talk about monetary arrangements or even reveal the name of the field holder. Eventually I learned which field belonged to which household and who the most frequent paid labourers were, but initially this was a way to conceal the fact that labour sometimes was hired in the village. It was much easier to start a conversation about farm labour exchange. In such conversations, both men and women often praised the fun and the emotional values of exchanging labour. On much fewer occasions did farmers openly admit to support the occurrence of paid labour.

Lan, my landlady, was frequently pointed out as very lucky, because her husband had an income as a teacher and both her sons earned money from working in big cities, which meant that they could sometimes contribute some cash. From time to time Lan used this money to hire farm labour. She herself said that as the only farmer in her household she was too busy to exchange much labour and she liked to organise things according to her own mind and be fairly independent. This, she carefully stressed, was to the advantage of the entire household. The more efficiently she organised the farm labour, the more time did she have for improving the conditions of the household, she reasoned. Even though she stated her opinion by using a discourse of household well-being and efficiency – qualities that the previous chapter identified as distinctive features of capable wives – it was subtly indicated to me by some villagers that her 'efficient' and business-like ways reflected a certain level of coldness and dissociation. As pointed out earlier, both Lan's efficiency and her wish to organise her work to support her family and facilitate her husbands' professional career could find ideological support at local as well as national level (Ngo Thi Ngan Binh 2004: 62–68; Tran Thi Van Anh and Le Ngoc Hung 1997: 87–89). Nonetheless, the fact that she did not always prioritise labour exchange and instead hired comparatively much labour, provoked comments from her fellow villagers. Her focus on the family's well-being, and all the farm work that she engaged in herself, did not fully outweigh the way she made use of money to access farm labour. It was relatively accepted for women to hire labour when they were overburdened with work, or if they for some reason were prevented from participating in labour exchange. But there was a fine balance between hiring enough labour to cope with immediate demands, and paying others to work because you could afford it and, thus, consciously decrease your participation in labour exchange.

In order to reduce the provocative effects of hired labour, Lan mainly hired women and men from the neighbouring village, where she and her husband had lived when their children were small. Even if people still noticed that she hired labour, it was less conspicuous than to hire fellow villagers. People from other places, 'outsiders' in other words, were not really included in the local community and this made it easier to avoid unbalanced relations within the village.

Another woman who challenged the idea of preserving village community through women's farm labour was Thu. She was from the southern part of the country and had married a farmer from this village. Her situation as a woman in her 30s without formal employment and with two small children was not at all unusual. What was unusual was her way of managing her farming situation. Normally, when a woman married a farmer, she spent a large amount of time cultivating the household's fields, and often engaged extensively in farm labour exchange with for instance sisters-in-law. On one occasion, when we were alone in her home, which she also shared with her mother-in-law, Thu told me that she hired all the farm labour that her household needed. She had married into a comparatively business-minded and affluent family – her husband was one of five relatively industrious brothers – and therefore she could afford to hire labour. The fact that Thu did not have any relatives of her own in this area certainly had an impact on her choice. She had no lingering labour obligations as a daughter or a sister, to consider. A woman of lesser means and greater kin network in the village would have found it very difficult to totally renounce farm labour. Her decision not to practice farming herself, and not to engage in farm labour exchange, positioned her outside these female village networks. Her social relationships consisted largely of her in-laws and she was rarely seen out and about in the village. But her somewhat detached position, as far as the community was concerned, arose from her and her in-laws' relatively affluent situation, not from economic vulnerability – and not really from social frailness either. Her withdrawal from farm work implied an independence from the farm-based village communality. She and her household, as well as the households of her brothers-in-law, employed an alternative, more 'progressive', way of making a living, where farming was not the main occupation and where labour relations with kin and co-villagers were not necessarily the most important.

However, for the continuance of an atmosphere of village community, it was important *how* farm work – and especially women's labour – was organised and compensated for. A small group of women, married as well as single, were regularly seen working in the fields of fellow villagers for cash. This was a way of gaining an additional income to their own farming. In an attempt to somewhat professionalise their work a few of these women expressed a certain pride in being skilful and appreciated farm labourers whom people wanted to hire. However, in contrast for instance to men's professionalised ploughing, paying for these women's farm labour inescapably added an element of inequality to the agreement. Some of these women were, for example, often the last ones to get their rice harvested, because they had been busy hiring out their labour as much as they possibly could in peak season. Their labouring for payment in other villagers' fields revealed these women's relative poverty; the other villagers knew they could not afford to turn down work opportunities and they themselves never paid for farm labour. Differences between the households in the village thus became obvious. Even though both parties were in agreement, the idea of intra-village cooperation was challenged by direct payment for general farm labour.

Hoa, who engaged in labour exchange but also hired some labour when she needed to, said that according to her 'it is better for the emotions to exchange labour. If you only hire labour and never work on the fields yourself, you will seem like some kind of boss and people do not like this.' She expressed what some of the villagers feared – to be perceived as behaving like a landlord, i.e. someone who exploited other people's labour without engaging in farm work themselves. From these points of view, women's exchange of farm labour was a critical counterbalance to disloyalty and inequalities within the village. Therefore, the way they organised and compensated for farm labour was perceived also through a moral lens, with implications for themselves as well as for the community.

Men's 'outside' position in farm labour exchange

In the case of ploughing and harrowing, payment in cash frequently took place. Even relatives quite often gave ploughmen money for their labour, although often at a reduced price. Ploughing and harrowing were normally performed by men, who, more or less, had professionalised this line of work. These men owned buffalos, or in a few cases cows, as draught-animals, and had the necessary equipment to perform their skills. Ploughing and harrowing were, as pointed out in the preceding chapter, the farming chores most closely associated with men. Even though rice transplanting held a similar position for women, this was not acknowledged as a 'professionalised' chore that individual women could specialise in, and for which they normally would choose cash payment. In peak periods, the ploughmen's services were highly sought after and they worked from dusk till dawn in the fields. One reason why they preferred cash instead of labour as compensation for their work was that they were so busy during the periods when their labour was needed. It would have been too complicated to keep track of all the households who owed them, and they would have had more farm labour than needed if everyone repaid them in work. Ploughing was perceived as a limited and specialised service because it could only be performed by one man and his buffalo. This 'male' work could be bought for money without challenging the ideal of cooperation between relatively equal neighbours.

Even if men often talked about the benefits of labour exchange, and many of them enthusiastically adhered to the principle of mutual help and support within the village, they rarely participated in farm labour exchange themselves. When men did participate in other types of farm labour than ploughing and harrowing, such as carrying harvested rice, seedlings, and manure, their compensation rather often consisted of a meal provided by the field holder. This normally took the form of a dinner served by the woman in the household that had received these men's assistance. Her husband who himself might, or might not, have participated in carrying things, acted as the host and shared the food with the other men. It was not always the case that the woman, who most likely had harvested or transplanted the field, and cooked the meal, joined for the dinner. The meals, which always had to include generous amounts of alcohol, were largely a male affair and integrated neatly into a male-centred community where farming did not constitute the centre. These meals were a way to emphasise friendliness or kinship ties between men, rather than formal labour agreements. As the previous chapter discussed, men also had an interest in expressing their participation in farm labour in terms of voluntariness and fun (*vui*), or pleasure, when it was not a question of more professionalised services. Women, on the other hand, were not customarily compensated for their farm work with a festive lunch or dinner with accompanying liquor. Men's labour compensation – often in cash or in the form of a dinner party – served to separate men from the female-dominated networks of labour exchange (Carsten 1989; Moore 1988: 61; Strathern 1988: 96). In practice, it also detached these men from the 'emotional' and non-hierarchical intra-village relations that the concept of *tinh cam* found support for in the realm of exchange of farm labour.

Spatial segregation of farm labour

As indicated by the examples above, there was an element of spatial segregation incorporated into the arrangements when ploughmen worked by themselves and were paid in cash for their labour, or when men who had been carriers were compensated with a dinner. Gendered division of farm labour often, in itself, entailed spatial segregation, when men and women performed different duties at different times and places. It did, however, happen that men and women exchanged farm labour with each other, both as kin and as neighbours. Since they, in many respects, worked with complementary tasks they, of course, needed access to each other's areas of work and found varying ways to come to suitable agreements. When men and women formed working relations, as the introductory example showed, the moral aspects were not only concerned with fairness and village sentiments, but also with the virtue and reputation of the parties concerned.

When Thuyet, a woman in her forties, explained to me how she organised her farm work, it was clear that men and women were not interchangeable as working partners. She thought that women normally felt most at ease when working in the company of other women: 'Women work at the same pace and are more or less equally strong and they chat and laugh well together. You do not do this with men.' The women Thuyet referred to were primarily other farming women, kin and non-kin, from Lang Xanh. She engaged in labour exchange with many of them, but also used hired labour if time was short. Sometimes she also exchanged labour with men, but such exchanges implied that they would settle the scores by performing separate duties and, quite likely, at different times and places. Thuyet explained how labour exchange with women and men could differ:

> For example, I might transplant rice for one woman and later maybe she helps me to plant manioc. But we will do both of these things together. If it is a man, I could ask him to carry manure to my field and then I will transplant for him. I would not carry manure together with him and he would not transplant with me.

Thus, temporal and spatial separation as well as task-oriented separation, all governed by perceptions of gender, was taken into consideration and then applied in each situation. A man who, for example, joined a group of transplanting women would thus be quite 'out of place' as the fields were women's domains when female tasks were performed (Ahmed 2006: 135–36; McDowell 1999: 145). The 'stiff'and 'hard' male body was not oriented toward this type of work, and could not inhabit a field that was in the midst of being transplanted, with the same 'naturalness' as the female bodies did. The body of a transplanting man would not have a given place in the field. If we follow Ahmed's reasoning, this was so because his movements would not have been tuned in with the task and the environment (Ahmed 2006: 62). If men were present at all, they usually operated in the fringes of the female group, where they performed carrying and transportation tasks.

One day, when I found Hanh (the woman from the introductory example) – this time together with her friend Toan – working in a field, she again drew my attention to the moral implications of disregarding the gender aspects of compensational agreements. The two women transplanted a smaller field for a man who had married one of Lai and Bao's daughters; in other words, a man who had not brought a house or residential land into the marriage and who's wife was not a farmer but a fulltime teacher. He himself was not present. With no female farmer in the household he was quite disconnected from the networks of labour exchange. Already at an earlier stage, he had explained to me that he wanted to give prominence to his work as a carpenter, even if his assignments were relatively few. And without a cow or a buffalo, he had little 'male' labour to offer in return for Hanh and Toan's work. Thus, the only way he could compensate them was in cash. There was, of course, the option that they help him to transplant the field for free, Hanh suggested, but it would then have indicated a much more intimate relationship. In this case, compensation in cash was therefore important: 'If not, people might wonder what kind of relation I have with him', Thanh giggling but firmly informed me. The moral and spatial demarcations between male and female labour and labourers implied a certain risk if a man and a woman were to engage in labour exchange by working together with the same task in the same place. Both of them would be 'misplaced' when neither spatial nor task separation was practiced, and the field would lose its meaning as a morally 'safe' place, especially for women.

Feminisation of farming – expanding the 'inside'

Gender-separated farm work was supported from a more historical perspective, where in the past good female conduct and morality, as far as possible, prescribed a gender segregated labour organisation, where women's place was secluded to the 'inside'. Looked at from this angle, the association between a spatially demarcated female labour realm and the 'inside' can be traced back to Confucian morals, where Vietnamese women's work traditionally was confined to the 'inside' of the household. It was within the protective walls of the house, or in the compound surrounding it, that the dignity and morals of the female household members could best be safeguarded (Ngo Thi Ngan Binh 2004: 50). In Vietnam's rural areas throughout history, women have taken an active part in farm labour, and the segregation of male and female labourers has not always been strictly enforced, or even practically applicable, due to harsh circumstances and extensive labour needs (Luong 1992: 72–73; Pham Van Bich 1999: 35–36). Nonetheless, separation of male and female farm labour – in tasks, compensation, and place – is still having implications for the respectability and moral status of the farmers in Lang Xanh.

The contrast between 'outside' (*ngoai*) and 'inside' (*noi*) was often used when people in Lang Xanh talked about different work and referred, for example, to spatial, compensatory, and gendered aspects of labour. [2] With an increasing number of (still mostly male) villagers engaged in

2 As mentioned already, the analogy of 'inside' and 'outside' was used in other contexts as well. In chapter 3 we encountered theses concepts in the realm of paternal and maternal kin groups, and when referring to for instance villagers and non-villagers. Another example is the contrast between Vietnamese and non-Vietnamese. See for example Hy Van Luong 1992; Pham Van Bich 1998; Kleinen 1999; Hue-Tam Ho Tai 2001; Rydström 2003, for different aspects and uses of the contrast between 'inside' and 'outside'.

wage labour outside of Lang Xanh, new delineations of the places where work was performed was continuously going on. For some men and a few women, these places now encompassed countries like Malaysia and Taiwan. Others went to work in larger cities or distant provinces in Vietnam for longer or shorter periods of time. This was, of course, regarded as 'outside' work. But occupations with a regular income, even if it was in the local school or health care clinic, were also perceived as 'outside' work. In these cases, it was the comparison with the unpaid farm work that virtually all the village households engaged in to some extent, which gave the salaried jobs the epithet 'outside'.

The conceptions of 'inside' and 'outside' work were not rigid and the demarcations were not defined once and for all (Niranjana 2001: 110). However, at this point in time, many of the farming duties were perceived as 'inside' labour by the villagers. Some of the reasons for this were the low income-levels, the small-scale activity, the general lack of recognized professionalism, and, of course, the proximity of the fields to the home. Farming formed a small but relatively solid base for the households to rely upon, but if they wanted to increase their wealth, the Lang Xanh inhabitants had to search outside the village and the local farming community. As has been mentioned earlier, the distribution of land has remained rather equal and did not play any significant role in the socio-economic differences between the villagers. Incomes that held the potential to create greater divergences have been a result of other activities, outside household farming. It was clear that farming was not a lucrative profession, but more of a family concern where daily needs of sustenance were incorporated into everyday household labour routines. The perception of farming as largely a household concern was not at all new, but the marked contrast between this type of labour and an economy based on production, growth, education and proficiency, was unique for this period.

However, the domination of women definitely meant that the regular farm work, they did in the fields, increasingly was perceived as housework, and consequently performed on the 'inside'[3]. Women's

3 As a result of the earlier phases of economic reform in the 1990s, more women than men lost their work opportunities in the government section, when this was drastically slimmed to give room for a market-oriented economy with a private sector. When this happened, many women went back to farming on the pieces

main responsibility for the households' farming contributed to a normative delineation, where the households expanded into the fields or, conversely, the fields were further incorporated into the households. Under the existing circumstances of farming in Lang Xanh, the female 'inside' labour sphere of housework had been widened to encompass the fields and the work that was performed there. When farm work was predominantly performed by women who remained 'in place' in the village, while many of the men laboured 'outside', then the meaning and understanding of the position of farm work in the households as well as in the village was altered. The demarcation of farming as largely and increasingly a female sphere, made the fields into a place that female bodies inhabited with ease, especially when 'female' tasks were performed. Now it was women's daily farm work, in combination with the intense labour periods when larger groups of women gathered in the fields, that formed the most common and tangible activity in the village. The presence of labouring women was a corporeal feature of village life.

The men who earned most of their income from farm related activities were neither placed in the realm of 'inside' labour, nor in the realm of 'outside' labour. Even though these men, just like the majority of the women, were regularly engaged in farm work, their labour was not referred to as 'inside' work. They were not considered to perform an extension of housework in the fields. Despite the air of professionalism that surrounded the ploughmen's work, they were not regular salaried employees (like for instance teachers or factory workers), and nor were they the owners of a business, so they did not perform 'outside' work. Hence the most 'male' farm work of all fell between the spheres of 'outside' and 'inside' labour.

If possible, Lang Xanh men worked mainly in the 'outside' sphere, earning money for the household. Men who did not bring in a sufficient amount of money from 'outside' labour found themselves, in some sense, to be 'out of place' if they stayed in the village without engaging in work that could be categorised as lucrative to a certain extent. Men's opportunities for working as hired farm labourers were rather limited

---

of land that the households had received during the land allocation. More men ventured into private enterprises, often in urban areas, in a quest for money – something that the small land plots could not produce in any greater quantities (Werner 2002: 35–36).

and very much concentrated in the peaks of the farming seasons. Only smaller amounts of money could be earned this way and the implications of this for a man's status and prestige were, as already pointed out, quite negative. There were no male equivalents to the 'inside' female dominated labour communities that existed in accordance with the cycle of the farming year.

Women's preservation of a place-based community

Let us recall, from the introduction and chapter 3, Young's argument that the cultivation and, to an even larger extent, the preservation aspects of Heidegger's conception of dwelling have been overshadowed by the much more valued activity of building. Young argues that when men have completed the building phase, women step in to preserve what men have constructed. According to Young, the housework that women do is often, if not always, confined to the underrated realm of preservation. Importantly, however, the daily work of preservation also helps maintain and care for the daily lives of people who dwell together, both on an individual and a communal level (Young 2005: 127, 142–43, 145). From this perspective, the fields in Lang Xanh were maintained and preserved as materialised relationships in and between the local community and the households. Like other types of household work, the result of farm labour was relatively impermanent and temporary. It was seasonally recurring and needed to be constantly repeated, just like much of the daily life sustaining work. In their capacity as women being 'in place', they would be present in the village where they as the main organisers and executors of farm labour continuously would be in contact with each other, forming into groups of varying constellations of close and distant relatives, as well as friends and neighbours.

All the time the women spent exchanging stories, thoughts, and worries, as well as rejoicings in the fields often gave them an opportunity to inquire about past and future labour requirements and have spontaneous encounters. Friends and relatives met in the village or stopped by each other's homes, to ask if they were available for harvesting or transplanting on a certain day. Through these female networks each household managed to assemble a working party when such was required. As some women got older and, hopefully, could reduce their presence in the fields, young women developed their skills and responsibilities, and spent more time farming. When daughters and sisters married and moved to their husbands' home villages, sons and brothers introduced new women to the labour networks through their marriages. By building on and referring to past, present, and coming generations, these networks of labouring women managed to connect the present with the past as well as the future (ibid.: 143). And, as many female farmers gave evidence to, over time, companionships developed between women from different households and among some of them, intimate links of trust and support were established. This kind of mutual loyalty formed a sense of village community.

Community, morality, and farm labour

Men's relative liberty to participate in more commercially oriented labour contexts gave them complementary roles in the socio-economic networks of a household. Even in the realm of farm labour, men found it comparatively unproblematic to engage in and talk about waged farm work. On the occasions when money was used as payment, it was important to 'know how to pay people good and fair', like Tuan, a farmer in his 50s, expressed it. He explained that it was not at all difficult to find people willing to work if they were compensated fairly, and that he knew who was usually in need of cash. His approach was rational and efficient; some villagers needed cash and he offered a just sum for their labour, and everybody was satisfied with the agreement, according to him. It was, by far, more accepted for a man to hire farm labour than it was for a woman to do so. It was not, as discussed in the previous chapter, expected that men should involve themselves in farming as much as possible in order to prove themselves as loyal family members and moral individuals. In the existing notions of masculinity there was not much room for men to exchange farm labour, as most of it was considered to take place in a female realm. It can be claimed that men's more detached labour relationships found support in the state's assertion of efficient and progressive labour and it connected their households with the commercial economy – especially when husbands spent most of their time working as paid labourers in outside places. Their time was considered, to a greater extent than women's, as having a value that was measurable in money, as in the case of the ploughmen (c.f. Bourdieu 1990: 117).

Women's exchange of farm labour, on the other hand, is frequently assessed – locally as well as a nationally – in the context of household duties (Le Thi 2001: 106–07; Molyneux 1981: 198; Moore 1988: 108–09; Werner 2002: 34). We have also seen that women were assessed in terms of individual as well as communal morality. Perceived through the lens of the government's economic and developmental aspirations, women's engagement in exchange of farm labour was 'backwards' and inefficient (Hoang Ba Thinh; 1998: 11–12; Le Ngoc Van 1997: 31–34; Le Thi 2001: 109; To Duy Hop 1997: 16–19; Taylor 2007). It did not contribute to a more industrialised, professionalised, and efficient agricultural sector. However, from the perspective of family and household maintenance, women's communal labour provided a network of support and mutual assistance. It also facilitated men's opportunities to involve themselves in money-generating activities that benefitted economic goals, but held the potential to increase inequality between persons and households. Ideally, women's roles in farm labour, as the preservers of solidarity between kin and neighbours, somewhat reduced the contradictions between the commercial and the communal values (Shreeves 2002: 227–28). Monetary contributions were, of course, important for the households, but it was not sufficient to hold a community together. And to some extent, transactions of money served to undermine the feeling of *tinh cam,* as the short-term and unequal aspects of these arrangements set the villagers apart for each other.

Women's daily engagement in farming kept many of the women 'inside' the village, close to the children and the elderly, enabling them to cooperate and exchange labour and ideas with women from other households. This nurtured notions of 'traditional' values of communality and considerate relationships between people who dwelled in the same place – in accordance with how this was symbolised in the idealised image of the rural village. In this way, women's exchange of farm labour also resonated with state-sanctioned female values such as selflessness, kindness and to care about other people (Le Thi 2001: 5–6; Rydström 2003: 51–53). Images of women with a fierce desire for money, who used unscrupulous ways to expand their income, occurred in popular media and literature as examples of the unwanted effects of a capitalistic economy. When women, on the contrary, engaged in exchange of farm labour they positioned their families and themselves in the moral realm of village communality.

Conclusions

A gendering process

This study set out to explore how female and male farmers in Lang Xanh, through their involvement with farmland and residential land, shaped and constituted themselves and others as gendered dwellers. The aim was to study how two types of land, in combination with some of the activities that took place there, formed part of a gendering process.

One central starting point was the relation between action and place, i.e. how female and male gender were shaped and perceived through people's actions in particular places and how, conversely, places took on different meanings and were made available to the inhabitants depending on whether the actors were male or female. However, what gendered bodies tended to do in certain places was also influenced considerably by history and habitual practices (Ahmed 2006: 56). The work and deeds of previous generations, in the fields and on residential land, laid the ground for the activities of the present inhabitants. Activities in the sphere of residential land were, for instance, informed by the presence of ancestors and otherworldly beings. Practices performed in and around the homes manifested relations between male and female dwellers, the patrilineal family, and the 'other world'. Also practices performed in the fields were inseparable from ideas concerned with the histories of the village, kinship, and household relations. Besides, the land-related practices, as well as the actors, were acting in contexts that were profoundly enclosed by ideological discourses.

Let us recall, from the introduction, Ingold's analysis of Bruegel's painting *The Harvesters* and how the labourers and the landscape in which they work are drawn together by the activity. None of the participants work in isolation from the others, but operate in a shared social environment where their work takes place in accord with the others.

The actors and their actions and movements are tuned in with the surroundings and with their fellow inhabitants in a relational, or rhythmic, way. They are, as Ingold puts it, *with* the environment and each other (Ingold 2000: 201–07). If we were to enter into similar scenes of Lang Xanh, but with the purpose to see not only how the inhabitants had a relational connection to the environment, but also how 'maleness' and 'femaleness' emerged and implied different ways of acting in and relating to each other and to these places, we could add some more nuances to the scenes.

To start with, the 'emerging' of male and female farmers was effected by existing discourses concerned with gender-specific traits and qualities, both at state level and in the local context. For the inhabitants in Lang Xanh, land was heavily imbued with governmental, political, economic, social, and individual meanings. Their surrounding environment was a consequence of past and present actions and this informed the everyday activities on the land. 'The landscape is both medium *for* and outcome *of* action and previous histories of action' (Tilly 1994: 23 italics original) as Tilly puts it.

As chapter 1 showed, the three categories 'women', 'peasants', and land have all been subject to much ideological attention. In preparation and organisation of the 1945 revolution, as well as in the subsequent political and state directives and policies, there have been extensive concerns about the capacities and desired qualities and purposes of these three categories. The men and women in Lang Xanh were, of course, aware of and influenced by these official discourses.

For example, on one hand, rural, and implicitly poorer, women were still sometimes pictured by the state as rich in virtue of the traditional kind, because of their more limited exposure to capitalism and the temptations of urban life (Taylor 2007: 29; see also Nguyen-vo Thu-huong 2008). On the other hand, they were also pictured as lacking something with respect to refinement and modern knowledge, compared to urban women. Thus, we can see that women in Lang Xanh could fulfil certain aspects of feminine ideals – like 'genuinity', virtuousness, and industriousness – which enhanced neo-traditional values that were assumed to protect the nation from corrupt foreign morals, while still provide incitement for a market economy (Nguyen-vo Thu-huong 2008: 46–54). But it was not the labour performed by peasant women on their households' rice fields that would allow Vietnam to develop into a respectable nation on a global scale. In short, the idea of the 'simple' but honest, caring, and hard-working rural woman was officially and somewhat commercially endorsed as a symbol of Vietnamese values, while the actual manual labour that these women performed in the fields was perceived as 'backwards' and coarse.

This meant that the dwellers in the fields were not the most highly esteemed citizens, and the small-scale farming that they performed took place on land that the state controlled, which made labour the most controllable asset of the farmers. The turbulent history and the strong ideological discourses influenced the farming techniques, the division of labour, and the availability of farmland. In this manner, the composition of the farming landscape in Lang Xanh was also part of national history and politics.

If we imagine, for a while, a scene where the landscape is composed mainly of residential land holding the houses where the villagers lived, a new picture will emerge. As chapter 3 explored, in contemporary Lang Xanh, the villagers still maintained what Ingold calls a relational connection to residential land (Ingold 2000: 149). This land had followed a path where the formal state ownership had left room for traditionally informed habits of access, control, and inheritance to continue. Intrinsic to the endurance of the patrilineal family and patrilocality has been the continuing practicing of ancestor worship in the homes. This practice has contributed to the sustenance of the patrilineal family, and played a significant role in placing men and women in gendered relationships vis-à-vis residential land and houses.

In contrast to women, men could, in their own capacity, draw together worlds and beings into a meaningful place where their families dwelled. And, not least, they built houses on the land, which turned the place into a meaningful union between worlds and beings. Women were incorporated from their fathers' patrilineages into their husbands' and, through their connections with men, they gained access to houses and residential land as daughters and wives. Instead of land, they brought their preserving labour into the marriage. Women contributed to the maintenance of the patrilineal family by nurturing the male builders and the other dwellers in this and the other world. These practices made residential land available to men and women on the basis of differing premises that, in turn, meant that they related to these places either as permanent and providing men, or as more temporary and receiving women.

With farmland, on the other hand, the relation was relatively detached from ancestral and divine contact, even though an idea of mutual nurturing between farmers and fields existed. There, the rupture in the relationships between farmers and the fields had been much more severe due to the rigorous reforms and allocations implemented by the socialist state after 1954. Access to and control of arable land was supposed to be gender-equal and men as well as women were allocated land from the state, which had taken on the traditionally patriarchal role as the provider of land. This change, I argued, had been possible because the detachment process, where farmers and particular fields had been spiritually and practically separated, had been quite systematic and extensive. One of the effects of the land strategies of the Vietnamese communist regime was the creation of a fracture in the relation between peasants and land. The purpose of this was to replace 'traditionally' informed land relations with a socialist one, where land – and especially arable land – could not be held and accumulated as private property. Land was not to be used as a vehicle for power by anyone, except for the state. Besides disrupting the power that ownership and control over land resulted in, the intent of the socialist land organisation was to turn land into purely a means of sustenance and rational production (Malarney 2002: 49). Access to farmland in Lang Xanh had been 'degendered' and 'desacralised' and it was labour rather than social and kin relations that entitled the farmers to use the land. However, farm labour had not been subject to similar ideological attention and reforms. The division of labour, for instance, kept being far from gender-neutral and it is worth noting that the question of altering men's labour duties never arose. The farmers continued to relate to each other and to the fields in clearly gendered ways.

If we consider these preconditions, and imagine that a successor to Bruegel is coming to the village of Lang Xanh, then we may have the scene for a contemporary Vietnamese version of *The Harvesters*. The golden harvest would be rice instead of wheat and the division of labour would be practically the reverse – the women would be the harvesters, bending toward the ripe rice, sickles in hand, cutting the stalks and tying them into sheaves. The men would be on their way to or from the field, either carrying heavy burdens to a waiting cart, or empty-handed, heading back to collect more sheaves. If the sheaves had not started to pile up yet, some of the men might wait in the shade of some trees or pause by the cart before they continued their work.

As chapter 4 showed, many of the inhabitants of Lang Xanh saw their manual farm labour as the most basic way of life, infused with a notion of following what was dictated by the environment and the innate capacities of male and female bodies. Many farmers in Lang Xanh thought the gendered labour division that governed farming, especially rice cultivation, could change only if manual farm labour was given up. Without mechanisation, farming was a constituting activity that shaped their labour, their opportunities, their social positions, and their bodies.

As we have seen, farming – especially rice transplanting – had been incorporated into the making of 'good' women and appropriate femininity, while for men, farm labour activities had to be engaged in with more care and consideration. Lang Xanh women were oriented towards land mainly through the labour they performed and, as we have seen, the quality and amount of their work were carefully monitored and evaluated by others. Women continuously risked being perceived as not working enough. Men, on the other hand, were oriented towards land as 'directors' through their assumed ability to inherit, control, and 'manage' places. Men did not so much risk to be seen as working too little, but rather to be working with the wrong things, and in this way endanger their positions as heads of households. Every performance of land-related labour, or refrainment thereof, realised the prestige, standing, and morals of each man and woman. Men did not have to constantly guard their respectability and prove their decency by labouring in the fields. Instead, men needed to confirm their positions as heads of their households. Hence, when men adopted feminine body postures and performed women's farm work, they risked reducing their masculinity. Tasks like transplanting and harvesting were regarded as 'small' not because these chores did not require any particular skills but because it was women who performed them in order to acquire female morality and respectability, based on their positions in relation to men and kinship. For Lang Xanh men, respectability and status were expressed not only by the 'big' work they did, but also by the labour that female family members performed on account of patrilineal families and, consequently, by the 'female' farming duties that they, as men and family heads, did *not* have to do.

In this way, the particular orientations of men and women to farm labour, and how they incorporated skills that tuned their bodies either into straight, hard, and heavy-handed, or into nimble, soft, and bendable individuals, gave the farmers a 'feel' for masculine and feminine qualities and positions. Restrictions and possibilities, as well as their relative responsibilities and abilities as men and women, were realised in practice in the performance of gendered farm labour division. This was an on-going process, a constant gendering of the male and female farmers in relation to each other and the environment in which they acted. In the performance of farm labour, ideas about gender traits could not be separated from the actual tasks, but constituted a lived reality, informative of the 'fates' of male and female farmers.

If we return for a moment to the harvesting scene, the people involved did not, as Ingold pointed out, act in isolation from each other, and their movements connected them with the environment as well as with the fellow labourers but, then I may add, they did so in gender-specific ways. To be a female farmer who bends to cut the rice, or a male farmer who carries the sheaves are two different ways of orienting oneself to the landscape and to each other. This informs the actors as well as the patient onlooker about what it can mean to be a man or a woman in this particular place.

If we should stay on and keep looking around in the landscape, we would over time notice that the human presence in the fields mostly consisted of women. I have argued that women's extensive presence in the fields, and their engagement in farm labour, constituted a perceptible and evident manifestation of the values of communality and the network of relationships that still linked the villagers together. With the intensification of commercialism and an escalating need to be part of a financial and labour market where local farming was not sufficient for sustenance, the village community – and even family units – were in some sense dispersing. Through their farm labour – though it was not seen as progressive and advanced – women contributed with their presence and preservation to a continuation of village and family based community.

Through women's continuous care for the fields, their 'inside' farm labour reinforced the idea of a genuine and communal solidarity typical of the rural village. Such a notion existed in Lang Xanh in the form of anecdotes and as means for self-cultivation; while in a national discourse, the idea of the rural community existed rather in the form of nostalgia and idealised authenticity (see also Drummond and Thomas 2003; Schlecker 2005). From this perspective, women's cooperation with other female farmers signified a sincere and tradition-based rural life that could also be associated with the romanticised perception of the unspoilt rural village (c.f. Strathern 1988: 81–82, 87). This notion was locally cherished as a counterbalance to a 'modern' and 'efficient' urban life, where labour might be more lucrative, but where people had forgotten how to treat each other with 'emotion' (*tinh cam*) and instead focused on individualistic and profit-oriented goals. Thus, the preserving capacity of women's farm work brought a context of continuity and unity to a changing world.

As a consequence of incorporating farming into the 'inside', feminine, unsalaried housework, this type of labour was increasingly associated with women's preserving and sustaining house-based work. Because of this, women still tended to be, first and foremost, associated with the more traditional female responsibilities of mothering, nurturing, and sustenance. As Moore points out, it is easy to get stuck in gender stereotypes, where women are reduced to the performance of domestic duties for daily sustenance and where men are 'outside' earning money (Moore 1988: 108–09). In the case of Lang Xanh, women's farm work could not be diminished to an intra-household concern, even though their farming activities were referred to as both 'small' and 'inside'. On the contrary, through their engagement in farming, they constituted and preserved the Lang Xanh community on a day-to-day basis. From this perspective, women's performance of farm work not only positioned women in the sphere of unending 'inside' drudgery, but also served to preserve what was for many villagers a highly valued household and village-based community, that could only be maintained by active presence and not by monetary compensation.

As this book has argued, by women devoting themselves to their farming duties, they contributed to the preservation of the family unit and the village community, at the same time as they – by constantly engaging in self-cultivation – manifested themselves, their families, and the village as decent and durable.

Hence, in many of the land-related situations that this book has concerned itself with, men and women related to, and were differently affected by, the places they inhabited and acted in. This influenced their sense of identity and agency, and consequently their feeling of belonging to a particular place (c.f. McDowell 1996: 38).

Concluding remarks

This study has drawn on Heidegger's idea that the acts of cultivating, building and preserving are the fundamental modes in which humans dwell in the world (Heidegger 1971: 145) and on Young's attempt to revalue the (typically feminine) work of preservation, which she argues mainly takes place in the home and is generally overshadowed by men's building activities (Young 2005: 125). So, while Heidegger mainly develops his ideas around building and Young takes up preservation of a house/home as an issue, none of them dwell on cultivation. The ability to draw the world into a meaningful place that Heidegger ascribe to buildings, and the spatial manifestation of identity and belonging that, according to Young, is one of the basic qualities of a home and its preservation, I would claim the same for the fields and farm work as well. As we have seen in Lang Xanh, the work of tilling the land – mainly done by women – linked together the dwellers as individuals into social and moral constellations of households, lineages and village community, as well as providing each farmer with a context for cultivating their individual traits and morality. Studying farming and engagement with residential land in Lang Xanh, the making and expression of meaningful identity and belonging could certainly be found in cultivation, and in other land related work.

When people move around and do things in their daily lives, the textures and the characteristics of their environment are incorporated into the bodies and sculpt, or shape, their mobility, movements, and motor skills (Ingold 2011: 47). This is exactly what happened when the women and men of Lang Xanh worked the land. Acquired knowledge and understandings of the land provided the inhabitants with their capabilities for actions. The feeling of being familiar with a place entailed insights into what was meaningful, expected, and acceptable to do there; in this way 'the place acts dialectically so as to create the people who are of that place' (Tilly 1994: 26).

Men and women's bodily performances of for instance ancestor worship, house building, and farming turned what was understood as gendered characteristics into lived reality. When they acted as members of households, kin groups, and village community, women and men *were* the traits rather than representations of them. In these actual, placed practices, gendered persons come into being (c.f. e.g. Bourdieu 1977: 87–95; Jackson 1989: 128; Popenoe 2004: 155–56). 'Male' and 'female' knowledge of ancestor worship and farming, generated certain understandings of what was expected and appropriate – and therefore meaningful – to do in a particular place.

This calibration of bodies and environment gave people who were familiar with a place a 'muscular consciousness' (Ingold 2011: 47), or a 'knowledge in the hands' (Merleau-Ponty 1962: 166), that made movements into incorporated habits of being in the world. The continuous reiteration of a body's labour, for example, makes the work appear effortless and 'natural'. The movements involved in the labour seem immanent to the body and this, in turn, makes the actor inhabit the place where the work is performed with the ease that comes with 'naturalness'. Or, as Ahmed puts it, 'what we "do do" affects what we "can do"' (Ahmed 2006: 59). And, as the farmers in Lang Xanh gave evidence of, it was a *gendered* muscular knowledge that made them interact with the land in ways particular to male and female actors. For example, when a group of women transplanted a rice field, this place was a female domain where men, if they were to engage in the same activity, could appear awkward and clumsy, and where the atmosphere of female companionship and relaxed socialising risked being disturbed. Ploughing, on the other hand, turned the fields into a male sphere, where individual men worked and women did not have a given place. The ease with which a transplanting woman, for example, dwelled in a field prepared for this task, gave the presence of a feminine body an air of self-evidence. The male body appeared as 'naturally' uncomfortable and out of tune in the same environment.

As a complement to Ingold's relational view on the interplay between skilled (but genderless) dwellers and the environment, Young's thoughts on gender structures can be useful. Young suggests that gender structures bring about constraints and opportunities that inform people's actions and situate them in a relational way. Even though each person takes to their varying situations according to their own experiences, and shape their practices accordingly, there are certain gender patterns which bring about implications that the individual cannot shape or control (Young 2002: 419–20). By adopting a conception of gender structures in addition to the more phenomenological conception of lived and placed bodies, a bridge can be established between what might, on one side, appear as an unreflected adoption of motor skills by 'un-gendered' practitioners with a 'natural' division of labour where men just happens to be ploughing the fields, and young girls learn to transplant, and, on the other side, gender conditions that, at least to an 'outsider', can appear as severely restricting and sometimes even coercive in their preconceptions. In combination, though, the two viewpoints can help us get a broader perspective on how gendering form part of our daily practices. Ideologies and understandings concerning gender capacities and characteristics form structures that women and men relate to and, through their lived bodies, incorporate and make use of to varying degrees. The daily enactment of being a respectable woman involved carrying out all the hard work that a Lang Xanh woman was expected to do, and indeed many performed this with increasing skill. Thus this enactment made conceptions about femininity a corporeal and, in a way, 'natural', part of being a woman at this particular place and time. This does not have to be perceived only as repetitive and passive compliance to repressive structures (i.e. lack of agency), but can also be actively used for operating in the world depending on the circumstances available for the actors.

Agency, as Mahmood claims, must not consist merely of resistance, but can also be the manner and means by which people take action and realise themselves as respectable persons (Mahmood 2005: 148, 157). As this book has argued, within the sphere of farming and labour, women in Lang Xanh found an accessible, acknowledged, and morally sanctioned place for 'self-fashioning'. This self-shaping took place in negotiation with the gender ideologies that were available to them, both on a collective and on a more individual level (See also Alsop et al. 2002: 81; Mahmood 2001). For the female farmers, the cultivation of their morals and assumed female capacity of selfless concern and diligence took place every time they laboured the fields. In this way, the farming women instilled in themselves, and incorporated, respectability and virtue. The actual practicing of farm labour generated not only skilled actors, but also women with a cultivated morality. Thus, the bodily act of farm labour was attuned to recognised Vietnamese feminine virtuousness (c.f. Mahmood 2005: 136).

By looking at what kind of farm work female and male bodies engaged in, the importance of the body as a physical and material entity, as well as a centre for experience and agency, is underscored. The body in place is the vantage point from where we experience the world and where our consciousness appears in material as well as spatial form (Low 2003: 10). If our subjective relation to the world is formed from how we, as bodily beings, orient ourselves to it, then the capacities which men and women are perceived to embody become crucial for how we dwell.

The postures of male and female dwellers, the way they moved and worked in the fields, for example, were creating a world that in certain ways held complementing and contrasting possibilities for actions and approaches, depending on the gender of the labourers (c.f. Bourdieu 1977: 89–92; Ingold 2000: 354, 358–60; Young 2005: 30). In other words, by embodying the movements, manners, and qualities of farming women and men, Lang Xanh inhabitants disposed themselves as acknowledged feminine or masculine dwellers in this place. However, as is hopefully evident by now, when men and women worked the fields – or involved themselves with residential land – they did not do this on equal, or neutral, terms. Their positions as gendered beings situated them differently, which in turn influenced men and women's possibilities to act on and relate to these places (c.f. Tilley 1994: 26). From this perspective, any notion of gender-neutral dwellers appears rather impossible. Dwelling occurs all the time, it is how we live, and it is therefore reasonable to assume that male and female farmers, as gendered subjects in locations, to a considerable extent dwelled in gendered ways.

References

Abrami, Regina M. and Nolwen Henaff. 2004. 'The City and the Countryside: Economy, State and Socialist Legacies in the Vietnamese Labor Market'. In *Reaching for the Dream: Challenges of Sustainable Development in Vietnam*. Melanie Beresford and Angie Tran (eds). Copenhagen: NIAS Press.

Abramson, Allen. 2000. 'Mythical Land, Legal Boundaries: Wondering about Landscape and Other Tracts'. In *Land, Law and Environment: Mythical Land, Legal Boundaries*. Allen Abramson and Dimitrios Theodossopoulos (eds). London: Pluto Press.

Ahmed, Sara. 2006. *Queer Phenomenology: Orientations, Objects, Others*. Durham: Duke University Press.

Alsop, Rachel, Annette Fitzsimons, and Kathleen Lennon. 2002. *Theorizing Gender*. Malden, Mass: Blackwell.

Arendt, Hannah. 1998 [1958]. *The Human Condition*. Chicago, London: University of Chicago Press.

Bélanger, Danièle and Magali Barbieri (eds). 2009. 'Introduction: State, Families, and the Making of Transitions in Vietnam'. In *Reconfiguring Families in Contemporary Vietnam*. Stanford, California: Stanford University Press.

Bergman, Arlene Eisen. 1975. *Women of Vietnam*. San Francisco: People's Press.

Bourdieu, Pierre. 1977. *Outline of a Theory of Practice*. Cambridge: Cambridge University Press.

———. 1990. *The Logic of Practice*. Cambridge: Polity Press in association with Blackwell Publishers.

———. 2000. *Pascalian Meditations*. Cambridge: Polity Press.

Bradley, Harriet. 1989. *Men's Work, Women's Work*. Cambridge: Polity Press in association with Basil Blackwell.

Brocheux, Pierre. 1980. 'Vietnamese Communism and the Peasants: Analogy and Originality in Vietnamese Experience'. In *Vietnamese Communism in Comparative Perspective*. Ed. William S. Turley. Boulder, Colorado: Westview Press.

Carsten, Janet. 1989. 'Cooking Money: Gender and the Symbolic Transformation of Means of Exchange in a Malay Fishing Community'. In *Money and the Morality of Exchange*. Jonathan Parry and Maurice Bloch (eds). Cambridge: Cambridge University Press.

———. 2004. *After Kinship*. New York: Cambridge University Press.

Carsten, Janet and Stephen Hugh-Jones (eds). 1995. *About the House: Lévi-Strauss and Beyond*. Cambridge: Cambridge University Press.

Chanh Cong Phan. 1993. 'The Vietnamese Concept of the Human Souls and the Rituals of Birth and Death'. *Southeast Asian Journal of Social Science* 21(2): 159–98.

Chu Van Lam. 1993. '*Doi Moi* in Vietnamese Agriculture'. In *Reinventing Vietnamese Socialism:* Doi Moi *in Comparative Perspective*. William S. Turley and Mark Selden (eds). Boulder, Colorado: Westview Press.

Cresswell, Tim. 1996. *In Place/Out of Place: Geography, Ideology, and Transgression*. Minneapolis, London: University of Minnesota Press.

Croll, Elisabeth. 1995. *Changing Identities of Chinese Women: Rhetoric, Experience and Self-Perception in Twentieth-Century China*. London: Zed Books.

Crossley, Nick. 2001. 'The Phenomenological Habitus and its Construction'. *Theory and Society* 30: 81–120.

Dang Phong and Melanie Beresford. 1998. *Authority Relations and Economic Decision-Making in Vietnam*. Copenhagen: Nordic Institute of Asian Studies.

Douglas, Mary. 2003. 'The Consumer's Revolt'. In *The Consumption Reader*. David B. Clarke, Marcus A. Doel and Kate M.L. Housiaux (eds). London: Routledge.

Drummond, Lisa. 2004. 'The Modern "Vietnamese Woman": Socialization and Women's Magazines'. In *Gender Practices in Contemporary Vietnam*. Lisa Drummond and Helle Rydström (eds). Copenhagen: NIAS Press; Singapore: Singapore University Press.

Drummond, Lisa B.W. and Mandy Thomas (eds). 2003. 'Introduction'. In *Consuming Urban Culture*. London: RoutledgeCurzon.

Duiker, William J. 1995. *Vietnam: Revolution in Transition*. Boulder, Colorado: Westview Press.

Endres, Kirsten W. 1999. 'Images of Womanhood in Rural Vietnam and the Role of the Vietnamese Women's Union: A Microscopic Perspective'. In *Vietnamese Village in Transition: Background and Consequences of Reform Policies in Rural Vietnam*. Bernhard Dahm and Vincent J.H. Houben (eds). Department of Southeast Asian Studies, Passau University.

Eisen, Arlene. 1984. *Women and Revolution in Vietnam*. London: Zed Books.

Gal, Susan and Gail Kligman. 2000. *Reproducing Gender: Politics, Publics, and Everyday Life after Socialism*. Princeton, New Jersey: Princeton University Press.

Gammeltoft, Tine. 1999. *Women's Bodies, Women's Worries: Health and Family Planning in a Vietnamese Rural Community.* Richmond: Curzon.

———. 2001. 'Faithful, heroic, resourceful: Changing images of women in Vietnam'. In *Vietnamese Society in Transition: the daily politics of reform and change'.* John Kleinen (ed.). Amsterdam: Het Spinhuis.

Goodkind, Daniel. 1996. 'State Agendas, Local Sentiments: Vietnamese Wedding Practices amidst Socialist Transformations'. *Social Forces,* Vol. 74(2): 342–59.

Gudeman, Stephen and Alberto Rivera. 1990. *Conversations in Colombia.* Cambridge: Cambridge University Press.

Ha Thi Phuong Tien. 1997. 'Some Problems About land and Rural Women in the Stage of Restructuring'. *Vietnam Social Sciences* 2(58): 56–64.

Hamilton, Roy W. (ed.). 2003. *The Art of Rice: Sprit and Sustenance in Asia.* University of California, Los Angeles. Fowler Museum of Cultural History.

Heidegger, Martin. 1971. *Poetry, Language, Thought.* New York: HarperCollins.

Hickey, Gerald Cannon. 1964. *Village in Vietnam.* New Haven: Yale University Press.

Hoang Ba Thinh. 1998. 'Agricultural and Rural Industrialization and Women's Role'. *Vietnam Social Science* 5(67): 3–30.

Hue-Tam Ho Tai. 1992. *Radicalism and the Origins of the Vietnamese Revolution.* Cambridge, Massachusetts: Harvard University Press.

———. 2001. 'Faces of Remembrance and Forgetting'. In *The Country of Memory: Remaking the Past in Late Socialist Vietnam.* Hue-Tam ho Tai (ed.). Berkeley: University of California Press.

Humphrey, Caroline. 2002. *The Unmaking of Soviet Life: everyday economies after socialism.* Ithaca and London: Cornell University Press.

Ingold, Tim. 2000. *The Perception of the Environment: Essays in Livelihood, Dwelling and Skill.* London: Routledge.

———. 2011. *Being Alive: Essays On Movement, Knowledge and Description.* London, New York: Routledge.

Jacka, Tamara. 1997. *Women's Work in Rural China: Change and Continuity in an Era of Reform*. Cambridge: Cambridge University Press.

Jackson, Michael. 1989. *Paths Toward a Clearing: Radical Empiricism and Ethnographic Inquiry*. Bloomington: Indiana University Press.

Jamieson, Neil L. 1993. *Understanding Vietnam.* Berkeley: University of California Press.

Kaneff, Deema. 2002. 'The Shame and the Pride of Market Activity: Morality, Identity and Trading in Postsocialist Bulgaria'. In *Markets and Moralities: Ethnographies of Postsocialism.* Caroline Humphrey and Ruth Mandel (eds). Oxford: Berg.

Keesing, Roger M. 1981. *Cultural Anthropology: A Contemporary Perspective.* Harcourt Brace College Publishers.

Kerkvliet, Benedict J. Tria. 2005. *The Power of Everyday Politics: How Vietnamese Peasants Transformed National Policy.* Ithaca, New York: Cornell University Press.

Kleinen, John. 1999. *Facing the Future, Reviving the Past: A Study of Social Change in a Northern Vietnamese Village.* Singapore: Institute of Southeast Asian Studies.

Kwon, Heonik. 2008. *Ghosts of War in Vietnam.* Cambridge University Press.

Le Cao Doan. 1995. 'Agricultural Reforms in Vietnam in the 1980s'. In *Vietnam in a Changing World.* Irene Nörlund, Carolyn L. Gates and Vu Cao Dam (eds). Richmond, Surrey: Curzon Press.

Le Kha Phieu. 2001. *Vietnam Entering the 21st Century: Selected Speeches and Writings of the General Secretary of the Central Committee of the Communist Party of Vietnam.* Hanoi: The Gioi Publishers.

Le Ngoc Van. 1997. 'Labour Change in Rural Families – new Roles of Males and Females'. *Vietnam Social Sciences* 2(58): 21–34.

Le Thi. 2001. *Employment and Life of Vietnamese Women During Economic Transition.* Hanoi: The Gioi Publishers.

Le Thi Nham Tuyet (ed.). 2002. *The Images of the Vietnamese Woman in the New Millennium.* Hanoi: The Gioi Publishers.

Le Trong Cuc and A. Terry Rambo. 2001. *Bright Peaks, Dark Valleys: A Comparative Analysis of Environmental and Social Conditions and Development Trends in Five Communities in Vietnam's Northern Mountain Region.* Hanoi: The National Political Publishing House.

Lévi-Strauss, Claude. 1983. *The Way of the Masks* (trans. S. Modelski). London: Jonathan Cape.

Li, Tania Murray. 1998. 'Working Separately but Eating Together: Personhood, Property, and Power in Conjugal Relations'. *American Ethnologist,* Vol. 25(4): 675–94.

Liljeström, Rita, Eva Lindskog, Nguyen Van Ang, and Vuong Xuan Tinh. 1998. *Profit and Poverty in Rural Vietnam.* Richmond, Surrey: Curzon Press.

Ljunggren, Börje. 1997. 'Vietnam's Second Decade under *Doi Moi*: Emerging Contradictions in the Reform Process?' In *Vietnam: Reformation and Transformation: Conference Proceedings*. Björn Beckman, Eva Hansson and Lisa Román (eds). Stockholm: Center for Pacific Asian Studies.

Low, Setha M. 2003. 'Embodied Space(s): Anthropological Theories of Body, Space, and Culture'. *Space & Culture*, Vol. 6(1): 9–18.

Luong, Hy Van. 1989. 'Vietnamese Kinship: Structural Principles and the Socialist Transformation in Northern Vietnam'. *Journal of Asian Studies* 48(4): 741–56.

———. 1992. *Revolution in the Village: Tradition and Transformation in North Vietnam, 1925–1988*. Honolulu: University of Hawai'i Press.

———. 1993. 'Economic Reform and the Intensification of Rituals in Two North Vietnamese Villages, 1980–90'. In *The Challenge of Reform in Indochina*. Börje Ljunggren (ed.). Cambridge, Massachusetts: Harvard University Press.

——— (ed.). 2003. *Postwar Vietnam: Dynamics of a Transforming Society*. Lanham, Maryland: Rowman & Littlefield.

———. 2010. *Tradition, Revolution, and Market Economy in a North Vietnamese Village, 1925–2006*. Honolulu: University of Hawai'i Press.

Mahmood, Saba. 2001. 'Feminist Theory, Embodiment, and the Docile Agent: Some Reflections on the Egyptian Islamic Revival'. *Cultural Anthropology*, 16(2): 202–36.

———. 2005. *Politics of piety: the Islamic revival and the feminist subject*. Princeton: Princeton University Press.

Mai Thi Tu and Le Thi Nham Tuyet. 1978. *Women in Vietnam*. Hanoi: Foreign Languages Publishing House.

Malarney, Shaun Kingsley. 1996. 'The Limits of 'State Functionalism' and the Reconstruction of Funerary Ritual in Contemporary Northern Vietnam'. *American Ethnologist* 23(3): 540–60.

———. 1997. 'Culture, Virtue, and Political Transformation in Contemporary Northern Viet Nam'. *Journal of Asian Studies*, Vol. 56(4): 899–920.

———. 2002. *Culture, Ritual and Revolution in Vietnam*. London: Routledge-Curzon.

Marr, David. 1981. *Vietnamese Tradition on Trial, 1920–1945*. Berkeley: University of California Press.

McDowell, Linda. 1996. 'Spatializing feminism'. In *Body Space: Destabilizing Geographies of Gender and Sexuality*. Nancy Duncan (ed.). London: Routledge.

———. 1999. *Gender, Identity and Place: Understanding Feminist Geographies.* Cambridge: Polity Press.

McElwee, Pamela. 2007. 'From the Moral Economy to the World Economy: Revisiting Vietnamese Peasants in a Globalizing Era'. *Journal of Vietnamese Studies,* Vol. 2(2): 57–107.

Merleau-Ponty, Maurice. 2002 [1962]. *Phenomenology of Perception.* London: Routledge.

Moise, Edwin. 1983. *Land Reform in China and North Vietnam: Consolidating the Revolution at the Village Level.* Chapter Hill: University of North Carolina Press.

Molyneux, Maxine. 1981. 'Women in Socialist Societies: Problems of Theory and Practice'. In *Of Marriage and the Market: women's subordination in international perspective.* Kate Young, Carol Wolkowitz and Roslyn McCullagh (eds). London: CSE Books.

Moore, Henrietta. 1988. *Feminism and Anthropology.* Cambridge: Polity Press.

———. 1994. *A Passion for Difference.* Cambridge: Polity Press.

———. 1996. *Space, Text, and Gender: An Anthropological Study of the Marakwet of Kenya.* New York: Guilford Press.

National Human Development report. 2001. *Doi Moi and Human Development in Viet Nam.* National Centre for Social Sciences and Humanities. Hanoi: The Political Publishing House.

Ngo Thi Ngan Binh. 2004. 'The Confucian Four Feminine Virtues (*tu duc*): The Old Versus the New – *Ke thua* Versus *Phat huy*'. In *Gender Practices in Contemporary Vietnam.* Lisa Drummond and Helle Rydström (eds). Copenhagen: NIAS Press; Singapore: Singapore University Press.

Ngo Vinh Long. 1973. *Before the Revolution: The Vietnamese Peasants Under the French.* Cambridge, Massachusetts: The MIT Press.

Ngoc-Luu Nguyen. 1987. *Peasants, Party and Revolution: The Politics of Agrarian Transformation in Northern Vietnam, 1930–1975.* Amsterdam: University.

Nguyen Duc Nghinh. 1993. 'Land Distribution in Tu Liem District According to Land Registers'. In *The Traditional Village in Vietnam.* Hanoi: The Gioi Publishers.

Nguyen Duc Nhuan. 1983. 'The Contradictions of the Rationalization of Agricultural Space and Work in Vietnam'. *International Journal of Urban and Regional Research* 01.7, 3: 363–79.

Nguyen Khac Tung. 1993. 'The village: settlement of peasants in Northern Vietnam'. In *The Traditional Village in Vietnam.* Hanoi: The Gioi Publishers.

Nguyen Sinh Cuc. 1995. *Nong Nghiep Viet Nam 1945–1995* [Agriculture of Vietnam 1945–1995]. Hanoi: Nha Xuat Ban Thong Ke.

Nguyen Tu Chi. 1993. 'The Traditional Viet Village in Bac Bo: its Organizational Structure and Problem'. In *The Traditional Village in Vietnam*. Hanoi: The Gioi Publishers.

Nguyen Van Huyen. 1995. *The Ancient Civilization of Vietnam*. Hanoi: The Gioi Publishers.

Nguyen Van Suu. 2004. 'The Politics of Land: Inequality in Land Access and Local Conflicts in the Red River Delta since Decollectivization'. In *Social Inequality in Vietnam and the Challenges to Reform*. Philip Taylor (ed.). Singapore: Institute of Southeast Asian Studies.

Nguyen-vo Thu-huong. 2005. 'The Real and the True: Neo-liberalism and Gender Governance in Vietnam'. Paper presented for Nordic Gendering Asia Conference, Kungälv, Sweden.

———. 2008. *The Ironies of Freedom: Sex, Culture, And Neoliberal Governance in Vietnam*. Seattle, London: University of Washington Press.

Nhung Tuyet Tran. 2008. 'Gender, Property, and the "Autonomy Thesis" in Southeast Asia: The Endowment of Local Succession in Early Modern Vietnam'. *Journal of Asian Studies*, Vol. 67(1): 43–72.

Niranjana, Seemanthini. 2001. *Gender and Space: Femininity, Sexualization, and the Female Body*. New Delhi: SAGE.

Parry, Jonathan and Maurice Bloch (eds). 1989. 'Introduction: Money and the Morality of Exchange'. In *Money and the Morality of Exchange*. Cambridge: Cambridge University Press.

Perrotta, Louise. 2000. 'Politics, Confusion and Practice: Landownership and De-collectivisation in Ukraine'. In *Land, Law and Environment: Mythical Land, Legal Boundaries*. Allen Abramson and Dimitrios Theodossopoulos (eds). London: Pluto Press.

Pettus, Ashley. 2003. *Between Sacrifice and Desire: National Identity and the Governing of Femininity in Vietnam*. London: Routledge.

Pham Cao Duong. 1985. *Vietnamese Peasants under French Domination, 1861–1945*. Berkeley Calif.: Center for South and Southeast Asia Studies, University of California; Lanham, Maryland: University Press of America.

Pham Van Bich. 1999. *The Vietnamese Family in Change: The Case of the Red River Delta*. Richmend, Surrey: Curzon Press.

Phan Dai Doan and Nguyen Quang Ngoc. 2002. 'Relationship between village, family clan and traditional family in the Red river delta'. *Vietnamese Studies* No. 3(145): 36–52.

Pine, Frances. 1996. 'Naming the House and Naming the Land: Kinship and Social Groups in Highland Poland'. *JRAI* (n.s.) 2 (2): 443–59.

Popenoe, Rebecca. 2004. *Feeding desire: Fatness, Beauty, and Sexuality Among a Saharan people*. London: Routledge.

Rydström, Helle. 2003. *Embodying Morality: Growing Up in Rural Northern Vietnam*. Honolulu: University of Hawai'i Press.

———. 2004. 'Female and Male "Characters": Images of Identification and Self-Identification for Rural Vietnamese Children and Adolescents'. In *Gender Practices in Contemporary Vietnam*. Lisa Drummond and Helle Rydström (eds). Copenhagen: NIAS Press; Singapore: Singapore University Press.

Schlecker, Markus. 2005. 'Going Back a Long Way: "Home Place", Thrift and Temporal Orientations in Northern Vietnam.' *Journal of the Royal Anthropological Institute* (N.S.) 11: 509–26.

Schuler, Sidney Ruth, Hoang Tu Anh, Vu Song Ha, Tran Hung Minh, Bui Thi Thanh Mai and Pham vu Thien. 2006. 'Constructions of Gender in Vietnam: In Pursuit of the "Three Criteria". *Culture, Health & Sexuality*, September-October; 8(5): 383–94.

Scott, James C. 1976. *The Moral Economy of the Peasant: Rebellion and Subsistence in Southeast Asia*. New Haven: Yale University Press.

Shanin, Teodor (ed.). 1971. *Peasants and Peasant Societies: Selected Readings*. Harmondsworth: Penguin.

Shreeves, Rosamund. 2002. 'Broadening the Concept of Privatization: Gender and Development in Rural Kazakhstan'. In *Markets and Moralities: Ethnographies of Postsocialism*. Ruth Mandel and Caroline Humphrey (eds). Oxford, New York: Berg.

Soucy, Alexandre. 2012. *The Buddha Side: Gender, Power and Buddhist Practice in Vietnam*. University of Hawai'i Press.

Spain, Daphne. 1992. *Gendered Spaces*. Chapel Hill: University of North Carolina Press.

Strathern, Marilyn. 1988. *The Gender of the Gift: Problems with Women and Problems with Society in Melanesia*. Berkeley, California: University of California Press.

Taylor, Philip. 2007. 'Poor Policies, Wealthy Peasants: Alternative Trajectories of Rural Development in Vietnam'. *Journal of Vietnamese Studies*, Vol. 2(2): 3–56.

Taylor, Sandra C. 1999. *Vietnamese Women at War: Fighting for Ho Chi Minh and the Revolution*. Lawerence, Kansas: University Press of Kansas.

Tilley, Christopher. 1994. *A Phenomenology of Landscape: Places, Paths and Monuments.* Oxford: Berg.

To Duy Hop. 1997. 'Higher Capabilities for Rural Women to Increase Their Role in the Course of Renovation'. *Vietnam Social Sciences* 2(58): 14–20.

To Lan. 1993. 'On Communal Land in the Traditional Viet Village'. In *The Traditional Village in Vietnam.* Hanoi: The Gioi Publishers.

Tran Ngoc Angie. 2004. 'What's Women's Work? Male Negotiations and Gender Reproduction in the Vietnamese Garment Industry'. In *Gender Practices in Contemporary Vietnam.* Lisa Drummond and Helle Rydström (eds). Copenhagen: NIAS Press; Singapore: Singapore University Press.

Tran Thi Van Anh. 1999. 'Women and Rural Land in Vietnam'. In *Women's Rights to House and Land: China, Laos, Vietnam.* Irene Tinker and Gale Summerfield (eds). Boulder, Colorado: Lynne Rienner.

Tran Thi Van Anh and Le Ngoc Hung. 1997. *Women and Doi Moi in Vietnam.* Hanoi: Woman Publishing House.

Tran Thi Van Anh and Wanjiku Kaime-Atterhög. 2000. *Vietnam Women's Union: Promoting Gender Equality.* Stockholm: Sida.

Troung Thanh-Dam. 1997. 'Uncertain Horizon: The Women's Question in Vietnam Revisited'. In *Vietnam: Reform and Transformation: Conference Proceedings.* Björn Beckman, Eva Hansson and Lisa Román (eds). Stockholm: Center for Pacific Asian Studies.

Werner, Jayne. 2002. 'Gender, Household, and State: Renovation (*Doi Moi*) as Social Process in Viet Nam'. In *Gender, Household, State: Doi Moi in Viet Nam.* Jayne Werner and Danièle Bélanger (eds). Ithaca, New York: Southeast Asia Program, Cornell University.

———. 2009. *Gender, Household and State in Post-revolutionary Vietnam.* New York: Routledge.

White, Christine Pelzer. 1982. 'Socialist Transformation of Agriculture and Gender Relations: The Vietnamese Case'. *Bulletin* 13(4): 44–41. Sussex: Institute of Development Studies.

———. 1983. 'Mass Mobilisation and Ideological Transformation in the Vietnamese Land Reform Campaign'. *Journal of Contemporary Asia* 13: 74–90.

———. 1988. 'Socialist Transformation of Agriculture and Gender Relations: The Vietnamese Case'. In *Southeast Asia.* John G. Taylor and Andrew Turton (eds). Basingstoke: Macmillan Education.

Wiegersma, Nancy. 1988. *Vietnam: Peasant Land, Peasant Revolution: Patriarchy and Collectivity in the Rural Economy.* London: Macmillian Press.

Wolf, Eric R. 1966. *Peasants.* Prentice-Hall; Englewoods Cliffs.

Wolf, Margery. 1972. *Women and the Family in Rural Taiwan.* Stanford, California: University Press.

———. 1985. *Revolution Postponed: Women in Contemporary China.* Stanford, California: Stanford University Press.

Yan, Yunxiang. 2003. *Private Life under Socialism: Love, Intimacy, and Family Change in a Chinese Village.* Stanford, California: Stanford University Press.

Young, Iris Marion. 2002. 'Lived Body vs. Gender: Reflections on Social Structures and Subjectivity'. *Ratio* 15(4): 410–28.

———. 2005. *On Female Body Experience: "Throwing Like a Girl" and Other Essays.* New York: Oxford University Press.

Index

Abramson, Allen 126
agriculture. *See* farming
agricultural collectives 26, 35, 36, 37–8, 64, 65, 74, 75, 103, 104, 118, 119, 179
  and gendered attitudes toward 22, 36
  and gendered labour division 43, 163, 164
  see also collective farming; labour; labour division; labourers; land; peasants
Ahmed, Sara 9, 187, 203
am/duong 162
ancestors 86–9
  ancestor halls 115
  altar 86, 87, 88, 92, 98, 100, 108, 109, 113, 120, 123, 125
  Communist Party's attitude toward 115–16
  and houses 86, 87, 89, 90, 108–9, 122, 125–26
  and land 116–17, 119–20
  and men 87, 89–90, 92, 94, 95, 96–8, 100, 102–3, 119–20, 125, 197, 203
  offerings to 52–3, 80, 87–8, 87 n.1, 95, 95 n.8
  and place 85, 86, 88–91, 98–9, 102–3, 107–9, 113, 125–26, 176, 195, 203
  tombs of 89, 102, 115
  transformation to the other world 89–90
  and women 87, 89, 92, 93–4, 95–8, 99, 119–20, 124–25, 197, 203
  worship 85–91, 92–4, 95–8, 99, 113, 114, 116, 117, 119, 120, 122, 126, 197, 203
  see also deities; divinities; family; ghosts; graves; houses; land; other world; place; residential land; rituals; spiritual beings; souls; spiritual matters; superstition
animal husbandry 54, 158–59
Annam (central Vietnam) 32 n.8
anti-communist regime 23

Binh Dinh province 119
bodies
  farming 118, 130, 134–36, 143, 164–66, 190, 199, 200
  female 9, 10, 130, 134–36, 143, 157, 158, 160, 162, 164, 165, 190, 199, 205
  gendered 8, 9, 10, 136, 160, 162–63, 164, 165, 187, 195, 199, 202, 203, 204
  male 9, 10, 135–36, 157, 158, 160–61, 162, 164, 165, 187, 199, 205
  place 8, 9, 136, 165, 187, 202, 203, 204
  see also dwelling; labour; place; skill; sexuality
Bourdieu, Pierre 129
Bruegel, Pieter the Elder 7, 134, 135, 195, 198
Buddhism 80, 113
  and women 79, 79 n.14, 80, 97 n.9
Buddhist Association 79 n.14, 80

buildings/building construction. *See* houses
Bulgaria 118

cadastral officer 64–66, 67, 69, 70, 71, 109, 138, 153
cadres 20, 20 n.4, 48, 55, 58, 66, 74–5, 78, 79, 116, 140, 141
  in land reformation 33, 34
Cambodia 141
Catholic Church 33
Central New Guinea 165
China 26, 52 n.4, 149
  gender aspects of house building 104
  gender relations 119 n.18, 119 n.19, 138, 142, 148, 163
  mother–son relations 101
Cochinchina 32 n.8
collective farming 74, 75
  end of 11, 75–6
  and gender equality 22–3
  women's work in 24, 25
colonial
  regime. *See* French colonial regime
  resistance 11, 17, 19, 29, 99 n.12
Confucianism 17, 29, 32
  family 17
  ideology 17, 162, 172
  village structure 31
  women/gender 18, 21, 23, 25, 44, 162, 188
communal house. See *dinh*
communal land. *See* land
communist ideology 35, 117–18, 161, 172, 179
  and land 31–9, 41, 115, 117, 127, 197, 198
  and peasants 30–41
  and women 21, 22–3, 25, 42–3, 44, 163, 196
Communist Party 20, 20 n.3, 23, 26, 35, 37, 40, 55, 56, 58, 59, 64, 74, 78, 115, 117, 161
Communal People's Council 55
cooperatives. *See* farming cooperative
councils of notables 29, 31
Cresswell, Tim 10
Croll, Elizabeth 101,104, 119 n.18
crops, types of 49–50, 49 n.3, 52–4. *See also* rice
cultivation
  self-cultivation 147, 167, 201, 204
  theory of 84, 85, 126, 181, 191, 202
  of women's traits 4, 6, 144, 148, 166, 167, 204.
  see also dwelling; human–environment relations; labour; morality; place
cultural house 56, 80–1
cultural village 61, 80, 140

daughters-in-law. *See* family
death
  anniversaries. *See* rituals
  preferred and non-preferred circumstances 89

Decree 10 38, 64 n.7
Decree 100 64 n.6
deities 52, 92, 96, 124
  and land 41, 85, 96, 109, 114, 116
  see also ancestors; divinities; ghosts; land; other world; place; residential land; spiritual beings; souls; superstition
Democratic Republic of Vietnam 21, 23
dinh (communal house) 77–9, 77 n.12, 80, 94, 113
District People's Council 55
divinities 80, 90
  divine relations 109, 113, 198, 161 n.5
  see also ancestors; deities; ghosts; land; other world; residential land; spiritual beings; souls; superstition
division of labour. *See* labour division

Doi moi (renovation, economic reformation) 11, 26, 38, 41, 43, 178
Douglas, Mary 178
Duiker, William J 32
duong. See *am/duong*
dwelling
  and fields 1, 6, 8, , 83, 193, 197, 203, 205
  gender aspects of 6, 85, 101, 109, 124–26, 193, 195, 197, 203, 202, 205
  and houses 83, 84, 94, 97, 109, 125–26
  theory of 6, 7, 83–4, 125–26, 191
  see also environment; human–environment relation; place

education 22, 25, 40, 61, 106, 132, 189. *See also* Lang Xanh
Eisen, Arlene 21
elite
  in anti-colonialism 29
  and communist revolution 34, 115
  patriarchal 120
  in village rule 29, 32
  see also intellectuals; literati
environment 6, 7, 8, 9, 41, 85, 114, 164, 166, 187, 196, 199, 200, 202, 203. *See also* dwelling; human–environment relation; place
equality
  between households 169, 178, 184, 185, 189, 193
  between women and men 21–2, 23, 42, 43, 44, 161, 198, 205
  and farm labour 22, 42, 161, 163, 172, 174, 178, 181, 184, 185
  of power 23

family
  Communist Party's ideal 26–7, 36, 115, 118
  Confucian ideal 17
  daughters-in-law 22, 23, 104, 106, 111, 132, 146, 147, 150, 151
  division of labour 42, 147, 164
  fathers-in-law 63, 100, 106, 108
  and men 98, 99, 105–7, 108–9, 123–26, 147, 150, 166, 167
  mothers-in-law 63, 67, 75, 78, 132, 146, 147, 150, 151, 181, 183
  obligations 121
  patrilineal 10, 98, 99, 100–1, 105–6, 108–9, 118, 120, 121, 123–26, 150, 167, 195, 197, 199, 202
  planning 56
  relations in rural areas 25
  son preference 89, 101, 105
  sons-in-law 106
  traditional structure of 22, 42, 118
  and women 98, 99, 100–1, 108–9, 119, 120, 123–26, 147, 148, 150, 164, 167
  see also ancestors; houses; households; 'inside'–'outside'; labour; labour division; labour exchange; land; Law of Marriage and Family; marriage; morality; patrilocality; place
Farm God 113, 114
farm labour, work. *See* labour
farmers. *See* peasants
Farmers' Union. *See* unions
farming
  community 171–72, 183–84, 191
  and deities/spirits 113–14
  output 37, 38, 39, 40, 114, 144
  seasons 2, 49, 51, 53, 54;
  and superstition 113–14
  techniques 41, 53, 197
  in warfare 36–7
  see also agricultural collectives; dwelling; family; households; labour; labour division; labour exchange; land; peasants; place
farming collectives. *See* agricultural collectives
farming cooperative (in Lang Xanh) 72, 73–4
farmland. *See* land
fate (*so phan*) 10, 30, 95, 108, 108 n.13, 117, 132, 151, 166, 200

fathers-in-law. *See* family
female-headed households. *See* households
femininity 11, 137, 140, 153, 166, 199, 200, 204, 205
  and cultivation 50, 199, 200
  state's ideas of 16, 44
  see also bodies; labour; morality; rural women; skill; women
feminisation of farming 5, 188–91
fields
  as gendered places 118–19, 127, 164–67, 186–88, 200
  irrigation of rice fields 72, 171
  manioc 49–50, 53, 68–9
  as meaningful places 84, 115, 117, 191, 202
  rice 50, 53, 64, 69, 73, 75
  size of 49
  see also dwelling; land; land access; land allocation; land law; land reforms; land use rights; place; residential land
fishponds 69–70
five goods movement 23
'five rights' 39
food
  as compensation for labour 185–86
  and gender aspects on socialising 82, 158
  rice, exceptional position of 53
  and rituals 87–8, 95, 95 n.8, 114
forest enterprises 50
fortune-telling 96, 97, 98, 107, 114, 115
four virtues 18, 23, 27, 146–47, 146 n.3
France, war against 15, 20 n.3, 23, 24, 59, 59 n.5, 76, 77, 99 n.12, 141
French colonial regime 15, 17, 19, 21, 29, 32, 33, 77
funerals. *See* rituals

Gammeltoft, Tine 146, 151
gendering processes 4, 5, 6, 9, 14, 129, 195–205
gender structures 10–11, 161, 203–4
geomancy 63, 97, 107
ghosts 87, 88, 89, 90, 93, 94, 113. *See also* ancestors; deities; divinities; other world; place; residential land; rituals; spiritual beings; souls; superstition
Górale 106, 122
graves 62, 63, 102, 113, 115, 119
graveyard (in Lang Xanh) 63, 89

Han Van Dong 78
Hanoi 16, 20, 40, 47, 48, 54, 55, 58, 63, 105, 151
Harvesters (painting by Bruegel) 7, analysis of 7–8
healthcare. *See* Lang Xanh
Heidegger, Martin 83–4, 85, 86, 109, 125, 191, 202
Ho Chi Minh 24, 78, 78 n.13
Ho Chi Minh City (Saigon) 27, 105, 147
home. *See* houses; place; residential land
household contract system 38
  and women's work 38–9
household garden plots 42–3, 121, 121 n.21
households
  classification of in land reformation 120 n.20
  'cultural' 140, 140 n.1
  economy 22, 26, 38–40, 42–3, 75, 137, 139, 177, 179, 189, 192
  endurance of 121–22
  female-headed 60–1
  heads of 22, 108, 114, 125, 137, 199
  labour organisation 38–9, 75–6, 137, 138, 139–40, 164, 166, 170, 172, 189
  landholdings after decollectivisation 39, 109–10, 111

residential structures 121–22; and women 27–8, 38–9, 42, 188
see also equality; family; houses; labour; labour compensation; labour division; labour exchange; village
houses
and ancestors 63, 85, 99, 100, 108–9, 122, 123–25, 197
building of 84, 91, 107–8, 125–26, 137, 191, 197, 202, 203
and collectivisation 120
endurance of 122
gender relations 86, 91, 99, 101, 105–9, 121, 197
improvements of 40, 47
inheritance of 90, 100, 102–3, 105, 115, 121, 122, 123
meaning of 84, 85–6, 109, 202
men's access to 90,100–6, 109, 121, 123–26
patrilineal connection 86, 94–5, 98–9, 101–6, 108, 115, 120, 125–26
and residential land 83, 85–6, 93–5, 97, 102–3, 121, 197
and spiritual beings 90–1, 93–5, 97, 108–9, 112–13, 122
women's access to 100–2, 106, 108–9, 121, 123–26
see also ancestors; dwelling; family; households; place; residential land; spiritual matters
human–environment relation 6–8, 9, 83–4, 85, 195–96, 199, 200, 202, 203. *See also* dwelling; environment; place

ICP. *See* Indochinese Communist Party
ideology. *See* Confucianism; Communist Party's ideology
Indochinese Communist Party (ICP) 19, 20 n.3,
inequality. *See* equality
Ingold, Tim 6, 7, 8, 9, 85, 86, 123, 165, 166, 195, 196, 197, 200, 203
inheritance. *See* houses; land; residential land; single women
'inside'–'outside' 89–90, 98, 98 n.11, 125, 150, 176, 188–90, 188 n.2, 193, 200, 201
intellectuals 17, 40. *See also* elite; literati
International Women's Day 140
Jackson, Michael 165
Jamieson, Niel L. 97 n.9

Kaneff, Deema 118
Kenya 139
Kerkvliet, Benedict J. Tria 36, 121
kinh (ethnic majority of Vietnam) 48, 48 n.1, 114
kinship. *See* family
Kleinen, John 113 n.16
Kwon, Heoink 89, 89 n.3

labour (agricultural) 1 n.1, 118, 123, 202
access to 22, 133, 140, 172, 182, 187
and bodies 129, 130, 134–36, 143, 146, 157, 160–61, 163, 164–66, 187, 189, 200, 203, 205
with buffalos/cows 2, 49, 133, 134, 137, 161, 170, 184, 185
changing attitudes toward 27, 29, 34, 41–2, 118, 176, 178, 179, 197
and community 169, 171–72, 173, 175, 177, 184, 191, 200
collectivisation of 22, 35, 37–8, 163–64
Communist regime's attitude to 41–2, 130, 163, 176, 178–79, 192, 193, 196–97, 198
gender aspects 106, 118, 123, 124, 130, 137–38, 139, 143, 144, 163
as an important asset 42, 112, 197
manual 160, 161, 177, 199
and men 5, 12, 22, 54, 57, 112, 129, 133–40, 142–44, 147,

151–59, 163–67, 170, 190, 199–200, 205
seasons 2, 8, 130, 132–33
skills 50, 133, 157, 161, 164–66, 170, 191, 200, 204
status of 34, 130, 141, 197
tasks 53, 130, 131, 132–3, 138–39, 143, 187, 188, 189, 199, 200
and women 27, 53, 54, 57, 112, 129–151, 155, 156, 163–67, 190, 191–93, 196–97, 199–201, 204–5
and young people 57, 132, 134, 138, 147, 150–51, 154, 164
see also agricultural collectives; bodies; family; houses; human–environment relation; labour division; labour exchange; labour organisation; naturalisation; place; peasants; rice; skill; work

labour compensations
and community 177–78, 183–84
gender aspects 173, 179, 180
hired/paid 51, 75, 169, 174, 177, 178, 179, 181–82, 193
and kin 174, 179–80, 181, 186
and men 170, 184–86, 187, 192
and non-kin 174, 179–80, 181, 186
and 'outsiders' 183
and women 170, 181–84, 186, 187, 192
labour division 8, 129, 197
and gender 1, 2, 9, 22, 43, 54, 85, 95–6, 97–8, 107, 129–67, 177, 186–88, 198, 200, 204
in farming collectives 24, 37, 43
and place 1, 110, 186–88
see also agricultural collectives; family; households; labour; labour organisation; place; rice
labour exchange 178, 36, 178, 180, 182
and community 177, 184
gender aspects 153, 170, 179–80, 186
and kinship 174, 179–80, 181, 186
and men 182, 184–86, 187, 192
and non-kin 174–75, 179–80, 181, 186
and women 181–82, 183, 187, 192–93
labour organisation 64, 74, 75, 130, 136–37, 171–72, 179
and friends 169, 175
gender perspective of 169, 179
and kin 169, 173–75
and men 169–71, 173, 175, 191
and neighbours 169, 174–75
and women 169–71, 173–75, 182
labourers
and access to farm land 127
in agricultural collectives 37–8
gendered 118–19, 188
lack of 140, 152, 153, 158–59, 173
recruitment of 172–75, 179–80
see also labour; peasants
land (arable) 11, 196
and ancestors 85, 89–90, 116–117, 119, 198
collectivisation of 22, 35, 74, 75–7, 117, 118, 119, 120, 121, 126
communal 23, 29, 31, 32
control of 11, 15, 31–3, 41, 77, 117–18, 127, 197, 198
desacralisation of 77, 78, 112–18, 126, 198
disputes 5, 71–3
during French colonial period 32–33, 32 n8
farmers' relation to 54, 83–4, 89–90, 117, 126–27, 160, 198
gender aspects of 4, 89–90, 110–11, 118–20, 127, 163, 166, 195, 198, 199, 203
importance of 50, 90, 117
inheritance of 18, 39, 50–1, 119, 120, 127
sacred/spiritual 41, 78, 112–114, 114 n.17, 116–17, 119, 126
as sensitive subject 58, 71–3
transactions 51, 65, 66–7, 68, 111–12, 126, 159

see also ancestors; agricultural collectives; residential land; place; rice
land access 33 n9, 64, 65–9, 70–2, 197
gender perspective 110–12, 119–20, 122, 170, 198
men 83, 110–12, 120
women 83, 110–12, 119–20
see also residential land
land allocation 11, 31, 38, 39–40, 42, 64–9, 64 n.7, 66 n.9, 76, 109, 109 n.14, 109 n.15, 110–12, 120, 122, 127, 179, 189, 198
discontentment with 70
Land God 90–1, 91 n.5, 92, 108, 109, 113. *See also* deities; houses; place; residential land; rituals
land law 39, 39 n.10, 40, 64, 65, 66, 66 n.9, 70, 109 n.15
gender aspects of 39, 120, 122
land reform 33–4, 74, 117, 120, 198
consequences of 34–5, 76–7, 78
errors in 34, 120–21
political purpose of 15–16, 33–5, 77, 78, 79
land use certificates (LUC/Red Books) 39, 66–7, 68
land use rights 39, 42, 109, 109 n.15, 127, 179, 198
landlords 32, 33–4, 35, 69, 75, 120, 178, 184
landscape 7, 8, 59, 85, 114 n.7, 195, 196, 197, 200
Lang Xanh
administrative/political structure 48, 55–8, 76
collective farming 65, 69, 74, 75, 76, 77
commune's police 57–8
as 'cultural village' 61, 140
education 62
fishponds 69–70
farming cooperative 72, 73–4
generational living 61, 63
graves/graveyard 63
health care 62
history 59, 77, 99 n.12
land allocations 64–9, 121
land conditions 49, 75
land issues 58, 71–3
land reform 74, 75
mass organisations (unions) 55–6
name 6, 76
patrilineages 59–60
physical landscape 47, 49–50, 75
population 59, 60, 62, 65
relation to neighbouring villages 48
residential land 68, 77, 120, 121
village guardian spirit 77, 78
village leader 58–9, 61–2, 71
village meetings 61–2
in wartime 59, 59 n.5, 60, 141
see also village
Laos 141
Law of Marriage and Family (1959) 21–2
Le code 119–20, 119 n.19
Le Kha Phieu 40
literati 28, 29. *See also* elite; intellectuals
lunar calendar 52 n.4, 53, 96, 107, 115
Luong, Hy Van 29, 33, 37, 79 n.14, 90, 120 n.20

Mahmood, Saba 204
maids, women working as 149, 158, 159
Malarney, Shaun Kingsley 95 n.8, 96, 97, 108 n.13, 114 n.17,
Malaysia 149, 189
Marakwet 139
market economy 11, 15, 26, 38, 40, 43, 171, 176, 178, 179, 189 n.3, 196, 200
marriage
equality in 21
and land access 122, 127
and men 103, 106, 123, 154, 187
and women 12, 27, 98, 99, 110, 125, 150, 151, 154, 192, 197
and women's labour 22, 147–48

Marxism–Leninism 30, 41, 179
Marxist historians 32
planned economy 26
masculinity 11, 44, 151, 153, 154, 155, 156, 164, 165, 192, 199, 200, 205. *See also* bodies; labour; men; skill
mass organisations. *See* unions
Mekong delta 49
men
and Confucianism 18
hot tempered 147–48, 154
and knowledge 8, 203
and laziness 145, 153, 161
in national/political ideology 24, 28, 44
relations to wife 137–39, 147, 153–54, 157
as veterans 154–55
in warfare 37
as weak 154–56, 157
work outside of farming 130, 137, 138, 147, 151, 155, 159, 187, 189 n.3, 190, 192, 193
see also ancestors; bodies; cultivation; dwelling; family; houses; households; labour; labour division; land; masculinity; mobility; residential land; sexuality; skill; superstition
mieu. *See* temple
mobility gender aspects of 2, 3, 82, 133, 149, 158, 159, 189, 190, 191–93, 202
Moore, Henrietta 129, 139, 201
Moores 158
morality
community 4, 179
Confucianism 29, 188
cooperative farms 37–8
farm work village community 140, 177–78, 193
men 152, 156, 169, 170, 186–86, 192, 202
political/national 23–4, 26 12, 144, 145, 169, 179, 202, 204
women 12, 25, 27, 44, 130, 140, 145–47, 148, 149, 150, 151, 156, 169, 184, 186–88, 193, 199, 202, 204, 205
see also cultivation; labour; place; men; women
motherhood 23, 27
mothers-in-law. *See* family

National Assembly 39, 55
National Day 56
naturalisation
of gender processes 4, 8–9, 164–67, 203, 204
labour division 159–62, 164–67
Nghe-Tinh uprising 30 n.7
Ngo Thi Ngan Binh 146
Nguyen Thi Lap
Nguyen Thi Minh Khai 20
Nguyen Tu Chi 90 n.4
Nhung Tuyet Tran 119–20

Old People's Union. *See* unions
other world (*the gioi khac*) 85, 89, 91, 95 n.8, 97, 115–16, 195, 197
otherworldly beings. *See* spiritual beings
'outside'. *See* 'inside'–'outside'

pagoda 79–80, 81, 95, 97 n.9
patrilineality 59, 115
patrilineal family. *See* family
patrilocality 41, 106, 110–11, 112, 121, 123, 126, 150, 197. *See also* houses; land; residential land
peasants 11 n.2, 28
access to land, in land reformation 35; ~ in traditional village 31–2
attitudes to collective farming 36–8, 118, 121
classification of 33, 33 n.9
in colonial resistance 28–9, 30
and Confucianism 29, 32
definition of 11, 11 n. 2, 15 n. 1
'The Peasant Question' 30, 30 n.6

in political ideology 15–16, 27, 29–30, 32, 34–6, 40–1, 42, 44, 118, 196
and reciprocity 29, 36
relation to fields 38, 41, 115, 118, 127
self-perception 41
superstition 31
revolts 30 n.7, 121
and worship 31
see also agricultural collectives; bodies; communist ideology; farming; households; labour; land
People's Committee 55, 56, 57, 59, 63, 79, 81, 89, 140
Pham Van Bich 89
Phu Tho province 36, 47, 48
Phu Tho town 59, 79
Pine, Frances 106, 122
place
ancestors 86–9, 94, 98–9, 115, 123, 197
community 89–90, 169, 176–78, 179, 191
Communist regime's control of 41, 115, 118
death and 89
gender aspects of 3–4, 6, 9, 84–5, 89–90, 101–5, 107, 118, 123–25, 129, 133, 143, 145, 150, 164–67, 186–92, 195, 196, 197, 199, 200–2, 203–5
ghosts/souls 88–9, 113
home- 89–90, 143, 176
Land God 90–1
morality 4, 13, 28, 144, 145, 149, 150, 186–92, 204
patrilineal family relations 115
socialising in Lang Xanh 81–2
spiritual beings 88, 90, 94, 97, 113, 115
theory of 6–8, 10, 117, 123, 165
see also ancestors; bodies; dwelling; environment; fate; houses; human–environment relations; labour; land; morality; naturalisation; residential land; village
Poland 106, 122
Popenoe, Rebecca 158
preservation
and men 100, 124
and women 13, 99, 100, 124, 125, 126, 148, 150, 169, 183, 191, 193, 197, 200, 201, 202

Red Books. *See* land use certificates
Red River delta 33, 48, 49
residential land
access to 68, 83, 103–4, 122–24, 197
and agricultural collective 103, 104, 121
and ancestors 85, 94, 102, 109, 115–17, 126
disputes 72
gender aspects 4, 11, 83, 86, 101–9, 121, 123–24, 195, 197, 202, 205
history of 94, 102–3, 112
and houses 83, 85–6, 94, 102–3, 107,108–9, 113, 121, 197
inheritance of 41, 100–3, 115, 121, 122, 123, 197
men 83, 101–9, 122–24, 127, 197
and spiritual beings 77, 78, 90, 93–4, 112–13, 115–17, 126, 197
women 83, 99–101, 105–6, 108–9, 122–24, 127, 197
see also ancestors; houses; households; land; patrilocality; place; spiritual beings; spiritual matters
revolution 1, 12, 15, 19
and control of land 31, 32, 83, 115, 196, 198
and men 19–20, 77, 98
and peasants 30, 40
and spiritual matters 29, 31
and women 19–21, 43, 74, 196
rice
and community of farmers 171–72
cultivation of 50, 53
importance of 50–4
lack of 51, 76

offering of 52–3, 80
as payment 51, 67
quality of 51
sign of affluence 51
see also crops; fields; food; labour; labour division; land; rituals
rituals 91–4, 117
agricultural 51, 52, 112–13, 113 n.16, 114–15,114 n.17, 116
communal 77
Communist Party's view of 52, 77, 115–16
death anniversaries 87, 87 n.1, 95, 95 n.8, 100, 105, 115, 117, 175
and exchange 107
festivals 62
funerals 62, 87, 88–9, 88 n.2, 107, 115, 142, 175
gender aspects of 97, 97 n.9, 97 n.10, 108–9
and house building 91, 108–9
new varieties of 41
weddings 62, 87, 107, 115, 137, 142
see also ancestors; deities; divinities; ghosts; other world; spiritual matters; spiritual beings; souls; superstition; *Tet*
rural women 11
and Confucianism 18
ideological ideas about 21, 25, 27–8, 44, 196–97
perceptions of 12, 28
see also single women; women
Rydström, Helle 10, 98 n.11

Sahara desert 158
secularisation. *See* land
self-cultivation. *See* cultivation
self-fashioning 154, 204
self-realisation 204 women's 11
sentiment. *See tinh cam*
sexuality 158
jokes about 58, 144, 157–58
men, 144 157–58
women 16, 144
single women
and ancestor worship 100
and houses 100–1, 108
inheritance 100
skill 6, 50, 157
gender aspects of 8, 107, 129, 147, 161, 164–66, 170, 191, 202, 203, 204
see also bodies; dwelling; labour; place
socialist states 16, 163
sons-in-law. *See* family
Soucy, Alexandre 79 n.14
Spirit of the Soil 113 n.16
spirit world. *See* other world
spiritual beings 53, 86, 87, 90, 93, 112, 113, 114, 115, 116, 144, 195. *See also* ancestors; deities; divinities; ghosts; houses; land; other world; place; residential land; rituals; souls
spiritual matters 52 n.4, 90, 113, 114, 123
and land 90, 112, 113, 116, 126, 198
and men 96, 97, 98, 108, 125
state's attitude toward 31, 41, 96, 115, 116
and women 79, 79 n.14, 96, 98, 125
see also ancestors; deities; divinities; ghosts; houses; land; other world; place; residential land; rituals; superstition
spiritual mediation 91–4, 96, 97
superstition
Communist Party's attitude toward 97, 113, 115
fear of 94, 96–8, 113
gender aspects of 94, 96–8
peasant–revolutionary opposition 31
see also ancestors; houses; land; place; residential land; rituals
souls 95
divine 88
ghost soul 89
wandering and placeless 88, 89, 94

see also ancestors; deities; divinities; ghosts; houses; land; other world; place; residential land; rituals; superstition

Taiwan 158, 159, 189
temple 112, 117, 120
  in the courtyard (*mieu*) 93, 93 n.6, 94, 96
Telefol 165
Tet (Vietnamese New Year) 51, 53, 64, 87 n.1, 95
Tet Doan Ngo 52
Thai Binh province 59
Thai Nguyen province 74
Thanh Ba district 47
'Three Capacities' 24
'Three Readinesses' 24
three responsibilities 141
three submissions 18
tinh cam (emotion/sentiment) 48, 172, 172 n.1, 175, 176, 177, 186, 193, 201
Tilly, Christopher 117, 126, 196
Tonkin (north Vietnam) 32 n.8
Truong Chinh 30, 30 n.6

UBND (*Uy Ban Nhan Dan*). *See* People's Committee
unions 55, 56
  Farmers' 55, 58
  Old People's 80
  Veterans' 56, 78, 80
  Women's 20–1, 22, 23, 25, 26, 27, 55, 56–7, 80, 138, 143, 170
  Youth 80–1
  see also Lang Xanh
United States
  support for anti-communist regime 23
  war against 15, 23, 24, 36, 37, 43, 59 n.5, 60, 141, 147

veterans. *See* men
Veterans' Union. *See* unions
village
  community/sentiment 169, 172, 175, 176–78, 183–84, 191, 192, 193, 195, 200–1, 202, 203
  men's ties to 89, 115
  restructured by communist revolution 34, 42, 43
  traditional land holdings 31–2, 33
  traditional structure of 29, 31–2, 33–4, 36
  see also Lang Xanh
village guardian spirit 77, 78, 97 n.9
Viet Tri 47
Vo Nguyen Giap 30, 30 n.6

weddings. *See* rituals
Werner, Jayne 21
White, Christine Pelzer 22
Wiegersma, Nancy 23, 90
Wolf, Margery 101, 138, 148
women 11
  as 'backward' 25, 193, 197
  knowledge 8, 203
  labour outside of farming 132, 137, 138, 147, 149, 149 n.4, 157, 158–59, 164, 189 n.3
  and laziness 145, 146
  in political ideologies 15–17, 20–1, 24–5, 26–7, 42, 43, 44, 140, 193, 196
  urban women's immorality 25
  in village power structure 31
  participation in warfare 24, 37, 140, 141
  as 'weak' 146, 157
  see also ancestors; bodies; Confucianism; cultivation; dwelling; family; femininity; houses; households; labour; labour division; land; mobility; morality; place; rural women; residential land; sexuality; skill; superstition; unions
Women's Union. *See* unions
work 1 n.1
  gendering 9, 11

outside of farming 4, 54, 130, 188–90
outside the home 188–90
outside the village 54, 62, 130, 188–90
see also bodies; dwelling; labour; men; peasants; place; skill; women

yin/yang. See *am/duong*
Young, Iris Marion 109, 125, 191, 202
Youth Union. *See* unions

zodiac
cycle 107–8
signs 94–5, 95 n.7